Japan's Engineering Ethics and Western Culture

Japan's Engineering Ethics and Western Culture

Social Status, Democracy, and Economic Globalization

Natsume Kenichi

LEXINGTON BOOKS
Lanham • Boulder • New York • London

Published by Lexington Books
An imprint of The Rowman & Littlefield Publishing Group, Inc.
4501 Forbes Boulevard, Suite 200, Lanham, Maryland 20706
www.rowman.com

86-90 Paul Street, London EC2A 4NE, United Kingdom

Copyright © 2021 The Rowman & Littlefield Publishing Group, Inc.

All rights reserved. No part of this book may be reproduced in any form or by any electronic or mechanical means, including information storage and retrieval systems, without written permission from the publisher, except by a reviewer who may quote passages in a review.

British Library Cataloguing in Publication Information Available

Library of Congress Cataloging-in-Publication Data

Names: Natsume, Kenichi, author.
Title: Japan's engineering ethics and Western culture : social status, democracy, and economic globalization / Kenichi Natsume.
Description: Lanham : Lexington Books, [2021] | Includes bibliographical references and index.
Identifiers: LCCN 2021020472 (print) | LCCN 2021020473 (ebook) | ISBN 9781793612892 (cloth) | ISBN 9781793612908 (ebook)
Subjects: LCSH: Engineering ethics—Japan—History. | Technology—Social aspects—Japan—History. | Japan—Relations—Western countries. | Western countries—Relations—Japan. | East and West.
Classification: LCC TA157 .N36 2021 (print) | LCC TA157 (ebook) | DDC 174/.962000952—dc23
LC record available at https://lccn.loc.gov/2021020472
LC ebook record available at https://lccn.loc.gov/2021020473

Contents

Acronyms and Abbreviations	vii
Acknowledgments	xi
Introduction	1
1 Engineering Ethics in Prewar Japan	9
2 Engineering Education and Ethics in Postwar Democratization	35
3 Import of the Western Engineering System and Its Ethics	61
4 Industry-Academia Cooperation: The Ideal and the Real	79
5 The Growth of Industrial and Practical Demands	97
6 The Globalization of Engineering Qualification and Ethics	127
7 The Globalization of Engineering Education and Ethics	147
Conclusion	175
Supplemental Glossary	181
Bibliography	191
Index	217
About the Author	227

Acronyms and Abbreviations

AAEEE	*Daigaku Denki Kyōkan Kyōgikai* (currently, *Daigaku Denkikei Kyōin Kyōgikai*); Association for the Advancement of Electrical Engineering Education
ABET	Accreditation Board for Engineering and Technology
ACES	*Kagaku Gijutsu Ōyō Rinri Kenkyūsho*; Applied Ethics Center for Engineering and Science of KIT
AICE	American Institute of Consulting Engineers
AIEE	American Institute of Electrical Engineers
AJCE	*Nihon Consulting Engineer Kyōkai*; Association of Japanese Consulting Engineers
APEC	Asia-Pacific Economic Cooperation
ASCE	American Society of Civil Engineers
ASEE	American Society for Engineering Education
ASME	American Society of Mechanical Engineers
CCS	Civil Communications Section of GHQ/SCAP
CE	Consulting Engineer
CI&E	Civil Information and Educational Section of GHQ/SCAP
CoCom	Coordinating Committee for Multilateral Export Control
CPD	Continuing Professional Development
CST	*Kagaku Gijutsu Kaigi*; Council for Science and Technology, *currently* Council for Science, Technology, and Innovation
EAJ	*Nihon Kōgaku Academy*; Engineering Academy of Japan
ECL	*Denki Tsūshin Kenkyūjo*; Electrical Communication Laboratory
ECPD	Engineers' Council for Professional Development, *currently* ABET

ENAA	*Engineering Shinkō Kyōkai*; Engineering Advancement Association of Japan
ETL	*Denki Shikenjo*; Electrotechnical Laboratory
FE	Fundamentals of Engineering
FIDIC	*Fédération Internationale des Ingénieurs-Conseils*; International Federation of Consulting Engineers
FJECIA	*Nihon Denki Tsūshin Kōgyō Rengōkai*; Federation of Japan Electric Communications Industrial Association
GHQ/SCAP	General Headquarters, Supreme Commander for the Allied Powers
IECEJ	*Denki Tsūshin Gakkai*; Institute of Electrical Communication Engineers of Japan, *currently* IEICE
IEICE	*Denshi Jōhō Tsūshin Gakkai*; Institute of Electronics, Information, and Communication Engineering
IEEE	Institute of Electrical and Electronics Engineers
IEEJ	*Denki Gakkai*; Institute of Electrical Engineers of Japan
IFIP	International Federation for Information Processing
ILO	International Labour Organization
IPEJ	*Nihon Gijutsushi-kai*; Institution of Professional Engineers, Japan
IPSJ	*Jōhōshori Gakkai*; Information Processing Society of Japan
JABEE	*Nihon Gijutsusha Kyōiku Nintei Kikō*; Japan Accreditation Board for Engineering Education
JCEA	*Nihon Gijutsushi-kai*; Japan Consulting Engineers Association, *currently* IPEJ
JCES	*Doboku Gakkai*; Japan Civil Engineering Society, *currently* JSCE
JEA	*Nihon Nōritsu Kyōkai*; Japan Efficiency Association, *currently* Japan Management Association
JFES	*Nihon Kōgakkai*; Japan Federation of Engineering Societies
JMA	*Nihon Ishi-kai*; Japan Medical Association
JMF	*Nihon Kikai Kogyō Rengōkai*; Japan Machinery Federation
JPC	*Nihon Seisansei Honbu*; Japan Productivity Center
JSCE	*Doboku Gakkai*; Japan Society of Civil Engineers
JSEE	*Nihon Kōgaku Kyōiku Kyōkai* (*Nihon Kōgyō Kyōiku Kyōkai*, until 1994); Japanese Society for Engineering Education
JSF	*Nihon Kagaku Gijutsu Shinkō Zaidan*; Japan Science Foundation
JSME	*Nihon Kikai Gakkai*; Japan Society of Mechanical Engineers
JSPS	*Nihon Gakujutsu Shinkōkai*; Japan Society for the Promotion of Science

JTTAS	*Nihon Kōgyō Gijutsu Shinkō Kyōkai*; Japan Technology Transfer Association
JUAA	*Daigaku Kijun Kyōkai*; Japan University Accreditation Association
JUSE	*Nihon Kagaku Gijutsu Renmei*; Japanese Union of Scientists and Engineers
KAKENHI	*Kagaku Kenkyūhi Hojokin*; Grants-in-Aid for Scientific Research
KIT	*Kanazawa Kōgyō Daigaku*; Kanazawa Institute of Technology
MCT	*Meiji Senmon Gakkō*; Meiji College of Technology
MEXT	*Monbu Kagakushō*; Ministry of Education, Culture, Sports, Science, and Technology
MITI	*Tsūshō Sangyōshō*; Ministry of International Trade and Industry, *currently* Ministry of Economy, Trade, and Industry
MOPT	*Teishinshō or Yūseishō*; Ministry of Posts and Telecommunications, or Ministry of Communications
MOE	*Monbushō*; Ministry of Education, *currently* MEXT
MOHW	*Kōseishō*; Ministry of Health and Welfare, *currently* Ministry of Health, Labour, and Welfare
NAE	National Academy of Engineering
NAS	National Academy of Sciences
NCEES	National Council of Examiners for Engineering and Surveying
NRC	National Research Council
NSF	National Science Foundation
NSPE	National Society of Professional Engineers
ODA	Official Development Assistance
OECD	Organisation for Economic Co-operation and Development
ONRI	*Orde van Nederlandse Raadgevende Ingenieurs*; Order of Netherland Consulting Engineers
PE	Professional Engineer
RCAST	*Sentan Kagaku Gijutsu Kenkyū Center*; Research Center for Advanced Science and Technology, the University of Tokyo
RHIT	Rose-Hulman Institute of Technology
RIITI	*Tsūshō Sangyō Chōsakai*; Research Institute of International Trade and Industry, *currently* Research Institute of Economy, Trade, and Industry
RIKEN	*Rikagaku Kenkyūsho*; Institute of Physical and Chemical Research

SCJ	*Nihon Gakujutsu Kaigi*; Science Council of Japan
SCM	*Gakujutsu Shingikai*; Science Council of the MOE
SEU	*Daigaku Secchi Kijun*; Standards for Establishment of Universities
STA	*Kagaku Gijutsuchō*; Science and Technology Agency
STS	Science, Technology, and Society
TEPCO	*Tokyo Denryoku*; Tokyo Electric Power Company Holdings
TTC	*Gijutsu Hon'yaku Center*; Technical Translation Center of the JCEA
UNESCO	United Nations Educational, Scientific, and Cultural Organization
WFEO	World Federation of Engineering Organizations
WMA	World Medical Association

In this book, in accordance with the seventeenth edition of *the Chicago Manual of Style* (University of Chicago Press 2017, 8.16, 11.89), the Japanese name is written so that the family name precedes the given name. In Japan, in alphabet writing, it has been customary for the given name to be written first, as it is in Western countries. However, in recent years, there has been an effort in Japan to conform to the original order in which names are written. This book, which discusses Japanese culture as a historical study, also follows this original order.

Acknowledgments

The astonishing accident at the Fukushima Daiichi Nuclear Power Station, caused by the Great East Japan Earthquake on March 11, 2011, was an immediate motivation to embark on this full-scale historical study of engineering ethics. Engineering ethics seemed to be an essential issue at the heart of the meltdown accident and warranted examination. Nevertheless, neither journalistic nor public debate forums provided adequate discussion points for the examination; thus, the lack of public awareness of engineering ethics became evident. Accordingly, Prof. Kihara Hidetoshi at Kokushikan University enthusiastically advised me to investigate the history of engineering ethics in Japan.

In 2006, I joined Kanazawa Institute of Technology (KIT) as one of the faculty members in charge of engineering ethics. Until then, I had never imagined I would be involved in engineering ethics education. I merely harbored the impression that engineering ethics was a new field related to science, technology, and society (STS). Although I was interested in STS, participated in related academic and civic activities, and even served as the representative of its networking organization (STS Network Japan); the history of electromagnetism in nineteenth-century Europe was my specialty. Meanwhile, I was rightly disposed to conducting historical research on engineering ethics because I was a specialist in the history of science and technology and belonged to a leading university in Japan for engineering education, especially engineering ethics education. Thus, the meltdown drove home the conviction that it was my professional duty to conduct this research project.

Many people supported this project to its completion. First, I would like to express my sincere gratitude to the following people who accepted my interview requests: Fudano Jun, Imai Kaneichiro,[1] Ishikawa Kenichi, Kobayashi Shinichi, Kusahara Katsuhide,[2] Murakami Yoichiro,[3] Nawa Kotaro,[4] Ohashi

Hideo,[5] the Institution of Professional Engineers, Japan (Fukumoto Muneki,[6] Hamada Tetsuo,[7] Hatakeyama Masaki,[8] and Kurosawa Takeo[9]), and the Japan PE/FE Examiners Council. Prof. Okita Yuji and Dr. Nakajima Hideto helped conduct the interviews. Although I am entirely responsible for the project, the respective support was vital to ground its credibility. I am also grateful to the Drucker Institute at Claremont Graduate University for granting permission to use their photographs as seen in figure 2.1.

I developed a better understanding of the historical subject via discussions at the Study Group for Contemporary Science and Technology Studies, hosted by Prof. Kihara, and the Japanese Society for Science and Technology Studies. In particular, Dr. Goto Ayako, a member of the study group, offered encouragement throughout the project. I also thank the KIT Applied Ethics Center for Engineering and Science (ACES) members and Dr. Tanaka Ichiro for stimulating the project through various activities. Dr. Okamoto Takuji steered me toward prewar higher education studies, especially regarding the case of Yamakawa Kenjiro. Dr. Elaine E. Englehardt read the final chapter and offered constructive feedback. The KIT Library Center staff have my gratitude for their support in locating and collecting research materials. The research was also conducted with the support of the Japanese Society for Engineering Education staff. I am indebted to many others for their advice and support; although I cannot list all names here, I nonetheless express my sincere gratitude to all of them.

Moreover, I thank the editors at Lexington Books for offering to publish this book at the Twenty-Eighth Annual International Conference of the Association for Practical and Professional Ethics (APPE) in Baltimore in 2019. Although I planned to publish my research results as a book, I had only thought of publishing it in the Japanese language for the Japanese domestic market. The possibility of an English publication with a worldwide reach became a great stimulus to develop the book to incorporate a broader and more sophisticated perspective. My friend, Dr. Shin Hiroki, also offered good advice on publication decisions and improvements.

This research was funded by the Kakiuchi Yoshinobu Memorial Award for Encouragement of Science and Technology Studies. Furthermore, chapters 1 and 2 were funded by JSPS KAKENHI Grant Number 17K04584. I also thank Editage (www.editage.com) for English language editing. I received financial support from the MEXT Private University Research Branding Project and KIT/ACES to use the editing service.

Regarding the contents of this book, the section on "Yamakawa Kenjiro's Bushido-Gentleman Moral Education" in chapter 1 is partially based on Natsume (2017b),[10] with additions on the topic of engineering ethics. Moreover, chapter 3 is based on Natsume (2014),[11] chapter 4 is based on Natsume (2016b) and Natsume (2017a),[12] chapter 6 is based on Natsume

(2015a),[13] and the "American Engineering Ethics Education at KIT" section in chapter 7 is based on Natsume (2016a).[14] I am grateful to the academic societies of the journals and the book publisher for their permission to employ the respective studies. I have made various revisions to the chapters based on the published papers and newly wrote the rest of the chapters.

Finally, I would like to thank my parents, Natsume Mikiya and Yumi, for their unconditional support throughout my career development, and my wife, a medical journal editor, Natsume Misaki, for all her encouraging support, from our private life to advice on writing this book.

The final stage of writing this book occurred during the COVID-19 pandemic. The situation made me reconsider how best to utilize science and technology to benefit society regarding democracy and economic activities. The future may present severer social problems regarding science and technology in a variety of forms that the world must address. Accordingly, engineering must continue to be a source of hope for humankind; its social expectations and responsibilities will become greater. This study provides an overview of not only engineering ethics but also its relationship with Japan's science, technology, industrial, and higher education policies. Moreover, it highlights basic national principles, such as nationalism and democracy. It is my sincere hope that this book will provide reliable and useful essential information to those interested in engineering ethics who yearn to better understand social issues on science and technology from a broad historical perspective.

NOTES

1. Imai Kaneichiro, interview by the author, June 23, 2012. Used with permission.
2. Kusahara Katsuhide, interview by the author, October 19, 2012. Used with permission.
3. Murakami Yoichiro, interview by the author, February 17, 2014. Used with permission.
4. Nawa Kotaro, interview by the author, August 22, 2012. Used with permission.
5. Ohashi Hideo, interview by the author, May 12, 2012, and Ohashi Hideo, interview by the author, August 4, 2014. Used with permission.
6. Fukumoto Muneki, interview by the author, July 13, 2012, and Fukumoto Muneki, email message to author, July 26, 2012. Used with permission.
7. Hamada Tetsuo, interview by the author, July 13, 2012. Used with permission.
8. Hatakeyama Masaki, interview by the author, August 31, 2012, and Hatakeyama Masaki, email message to author, September 1, 2012. Used with permission.
9. Kurosawa Takeo, interview by the author, July 13, 2012. Used with permission.
10. Natsume Kenichi, "Yamakawa Kenjiro no kagaku shisō to shōbu shugi: butsurigaku, shakaigaku, fukoku-kyōhei" [Yamakawa Kenjiro's Scientific Thought

and Pro-militarism: Physics, Sociology, and the Japanese Policy of Increasing Wealth and Military Power], in *Meiji-Taishō ki no kagaku shisōshi* [Essays on the History of Scientific Thoughts in Modern Japan: Japanese Thoughts about Science Approximately Between the 1860s and 1930s], edited by Kanamori Osamu (Tokyo: Keiso Shobo, 2017), chap. 2. Used with permission.

11. Natsume Kenichi, "Two Codes of Ethics Adopted by the Early Japan Consulting Engineers Association" [in Japanese], *Journal of the Japan Society for the History of Industrial Technology* 19, no. 1 (2014): 1–20. Used with permission.

12. Natsume Kenichi, "Engineering Professors and Industry-University Cooperation (1951–1969)" [in Japanese], *Journal of the Japan Society for the History of Industrial Technology* 20, no. 2 (2016): 11–19; and Natsume Kenichi, "Promotion and Criticism of Industry-University Cooperation in Japan from the 1950s to 1960s" [in Japanese], *Journal of Science and Technology Studies* 13 (2017): 32–47. Used with permission.

13. Natsume Kenichi, "Engineering Ethics and the International Consistency Problem of Gijutsushi in the 1990s: From Japan-U.S. Bilateralism to APEC Multilateralism" [in Japanese], *Journal of the Japan Society for the History of Industrial Technology* 19 no. 2 (2015): 17–41. Used with permission.

14. Natsume Kenichi, "Historical Details of the Development of Engineering Ethics Education at the Kanazawa Institute of Technology in the 1990s" [in Japanese], *Journal of the JSEE* 64 no. 1 (2016): 39–44. Used with permission.

Introduction

No one doubts if engineers who construct technological infrastructure in our society must be ethical. Therefore, should ethics be compulsory in engineering education? Otherwise, is ethics naturally emergent within the engineering community?

Globalization in the late 1990s paved the way for systematic ethics education (modeled on Western engineering education) to be introduced in Japan. Given the importance of ethics, was this introduction too late? Why was it introduced at that time, and why was it modeled on Western culture?

Notably, to register with the Asia-Pacific Economic Cooperation (APEC), engineers need an accredited professional qualification. Accordingly, to meet this requirement, Gijutsushi (a Japanese national qualification for engineers) transformed its qualification from consulting engineer (CE) to professional engineer (PE) in 2000. Furthermore, the Japan Consulting Engineers Association (JCEA, Nihon Gijutsushi-kai, currently the Institution of Professional Engineers, Japan) revised its code of ethics in 1999 and published a translation of an American textbook on engineering ethics in 1998, which became a pioneering textbook in Japan. The APEC requirement also encouraged Japan to establish the Japan Accreditation Board for Engineering Education (JABEE) in 1999, thereby making engineering ethics an essential requirement.

Moreover, many engineering professional and academic bodies in Japan have enacted codes of ethics since the late 1990s. Although it is necessary to distinguish between the histories of ethics education and ethics codes, only the Japan Civil Engineering Society (JCES, currently the Japan Society of Civil Engineers) and the JCEA had adopted professional codes of ethics before the 1990s. Table 0.1 charts the years when Japan's engineering societies enacted (or revised) their codes of ethics.

Table 0.1. Years of Enactment and Revision of Engineering Professional and Academic Bodies' Codes of Ethics in Japan

Year	Society
1938	Japan Civil Engineering Society (JCES)
1951	Japan Consulting Engineers Association (JCEA)
1961	Japan Consulting Engineers Association (JCEA)
1996	Information Processing Society of Japan (IPSJ)
1998	Institute of Electrical Engineers of Japan (IEEJ)
	Institute of Electronics, Information and Communication Engineering (IEICE)
1999	Japan Consulting Engineers Association (JCEA)
	Japan Society of Civil Engineers (JSCE)
	Architectural Institute of Japan (AIJ)
	Japan Society of Mechanical Engineers (JSME)
2000	Chemical Society of Japan (CSJ)
	Japan Society for Technology of Plasticity (JSTP)
	Illuminating Engineering Institute of Japan (IEIJ)

An increase in social demand could be another reason for introducing professional codes of ethics in the 1990s. Severe disasters and crimes regarding science and technology significantly impacted Japanese society in the late 1990s. Notable events include the Great Hanshin-Awaji Earthquake,[1] the sarin attack by the Aum Shinrikyo cult on the Tokyo subway,[2] the concealment of the sodium leak accident at the Monju nuclear reactor[3] (all of which occurred in 1995), and the criticality accident at the JCO's Tokaimura nuclear fuel-processing plant in 1999.[4] The ambiguity of scientists and engineers regarding their responsibilities was suspected as the major driver of these accidents and criticized as a negative aspect of Japanese traditional culture, such as collectivism and conformism. Therefore, engineering ethics in an individualistic culture are expected to provide a solution.

However, the introduction of engineering ethics education induced expectations, hesitation, and even rejection among Japanese engineers and engineering professors. They reasoned that it was from a strange culture driven by globalization, and Japan already had its traditional ways of nurturing the morality of engineers. Given the cultural heterogeneity, engineering ethics education has not been entirely accepted. Moreover, it has even lost the attention of the public.[5] In the aftermath of the Fukushima Daiichi nuclear disaster after the Great East Japan Earthquake in 2011, numerous discussions were held from the perspectives of science, technology, and society, with little mention of engineering ethics, except from those involved in engineering ethics education.

The fact that engineering ethics was not institutionalized in Japan until the 1990s does not mean such ethics had not been required in Japan before then.

Early examples in modern Japan regard Watanabe Hiromoto, who became a member of the House of Representatives after serving as the president of the Imperial University, and Fukuzawa Yukichi, Japan's leading intellectual on enlightenment. They emphasized engineering ethics in the 1890s (Fujie 2004, 159–161; Mikami 2006). Watanabe (1893, 4–12), in his 1892 speech at the Engineering Society (Kōgakkai, currently the Japan Federation of Engineering Societies), maintained that independent workers were responsible for everything they did, and engineering graduates, like politicians and physicians, had a weightier responsibility as a duty to their rights. He then blamed graduates for not gaining enough trust because they had a shallow sense of responsibility for their work; furthermore, they worked only for remuneration without considering their ability level. Fukuzawa ([1893] 1961a, [1895a] 1961b), in newspaper editorials in 1893 and 1895, also claimed that engineers had garnered a bad reputation. He blamed academic engineers from the old samurai class for failing to hold decency standards by forgetting to maintain their integrity after modernization and being guilty of dereliction of duty for personal gain, despite the necessity of trust in the engineering profession. The indispensability of engineering expertise to judge technological quality could not be overstated. Thus, he maintained that it was necessary to cultivate dignity in engineering education and generously increase the low financial rewards of engineers; moreover, the engineering profession should actively discuss their professional responsibilities (Fukuzawa [1895b] 1961c). Furthermore, Tejima Seiichi, principal of Tokyo Higher Technical School (currently Tokyo Institute of Technology), repeatedly questioned the character of Japanese engineers who mass produced shoddy products and took inspiration from Fukuzawa's argument for Japanese Bushido (the Samurai Way) and British gentlemanliness to promote character-building (Tejima 1909; Miyoshi 1999, 146–156). Moral education was compulsory in technical schools, not universities. However, it was part of the Japanese nationalistic education system, which was different from the ethics required in Western professional societies. Later, following the United States, the JCES enacted a code of ethics in 1938 from a global perspective that was never widely accepted. Chapter 1 discusses these prewar circumstances.

After World War II, the Allied Occupation of Japan marked further evidence of Western cultural influence. The Civil Information and Education Section (CI&E) of the General Headquarters, Supreme Commander for the Allied Powers (GHQ/SCAP), introduced a radical reform of the Japanese educational system toward antimilitarism and democracy. The democratization of education resulted in the transformation of many technical schools into universities. However, the reform reduced opportunities for vocational training. Thus, telecommunications engineering professors established the Association for the Advancement of Electrical Engineering Education

(AAEEE) in 1949 to improve the new system. As part of the reform, Koga Issac, an electrical engineering professor at the University of Tokyo and Tokyo Institute of Technology, promoted professional ethics education by recommending the American code of engineering ethics as a new educational guideline. Furthermore, the Japanese Society for Engineering Education (JSEE) was founded for all engineering disciplines in 1952. Moreover, its leader, Shimizu Kinji, also promoted engineering ethics. Unsurprisingly, American engineering ethics influenced these initiatives. Chapter 2 discusses the details.

The unique engineering ethics of the West (Downey et al. 2007; Mitcham 2019, chap. 12) had an unequal influence on Japan. Chapter 3 demonstrates that even though European countries had a significant influence on Japan's consulting engineers, the U.S.' influence was decisive, as it spearheaded the worldwide institutionalization of engineering ethics. According to the historiography of Mitcham (2001, 565–574, 2009, 2019, 155–168, 208–219), American engineering ideology can be divided into four periods. First, loyal obedience was emphasized in the late nineteenth and early twentieth centuries. Second, the pursuit and promotion of technical efficiency were emphasized in the first half of the twentieth century. Third, the protection of public safety, health, and welfare has been given prominence since the mid-twentieth century. Fourth, ethics education has been promoted since the 1970s. The professional code of conduct to define those values was initially implicit. However, the American Institute of Consulting Engineers (AICE) adopted a code of ethics in 1911; the American Institute of Electrical Engineers (AIEE), 1912; and the American Society of Civil Engineers (ASCE), together with the American Society of Mechanical Engineers (ASME), 1914. Around this time, as unions were promoted in the United States, several new engineering societies were established. During the Great Depression, the Engineers' Council for Professional Development (ECPD) was established in 1932 and began accreditation in 1934 to authorize engineering educational qualifications. Furthermore, the National Society of Professional Engineers (NSPE) was established in 1934 to license engineers and promote its legislation.[6] When the ECPD adopted the Faith of the Engineer in 1943 and the Canons of Ethics for Engineers in 1947, it became a model code of ethics for other engineering professional bodies regarding public safety, health, and welfare. NSPE also emphasized the integrity of professional practice from its inception and adopted the Canons of Ethics in 1946, ahead of the ECPD.

In the 1970s, accidents and misconduct continued, including the DC-10 airliner disasters, the San Francisco Bay Area Rapid Transit (BART) case, and the Watergate scandal. Accordingly, interest in ethics education increased in the United States. In 1974, the ECPD revised its code of ethics to make public safety, health, and welfare paramount. The Hastings Center, with support from the Rockefeller Brothers Fund and the Carnegie Corporation of New York, conducted a systematic study of ethics education in higher education from 1977 to 1979. Furthermore, in collaboration with academic engineers, philosophers, law experts, and social scientists, the Center for the Study of the Human Dimensions of Science and Technology at Rensselaer Polytechnic Institute and the Center for the Study of Ethics in the Professions at Illinois Institute of Technology launched the National Endowment for the Humanities (NEH) and the National Science Foundation (NSF) projects to pioneer engineering ethics courses and research (Baum 1980, v; Weil 1984, 341–343). Since then, the publication of textbooks on engineering ethics has progressed. In 1980, the ECPD changed its name to the Accreditation Board for Engineering and Technology (ABET). In the mid-1990s, it established Engineering Criteria 2000 (EC2000) to evaluate student outcomes through a flexible curriculum with engineering ethics as a requirement.

Unlike the United States, postwar Japan promoted engineering ethics education only in the late 1990s. Several studies on modern Japanese history have affirmed that economic globalization was the primary driver of engineering ethics. Given that its macroscopic perspective has been emphasized in the United States since the 1970s, Sugihara (2007) argued for integrating science, technology, and society (STS) studies with engineering ethics. Macroscopic arguments of social responsibility and corporate ethics have also been emphasized in Japan since the 2000s. Given that STS studies can integrate applied ethics in the environment, biology, and information research areas, the incorporation would have been seamless and simultaneous. Hiyagon (2015) also analyzed the reforms of the engineering qualification system and engineering education in the 1990s from three perspectives: the demands for ethics, human resources with problem-solving skills, and institutional preparation for globalization. He then illustrated how the rapid promotion of U.S.-modeled ethics had induced a lack of understanding, confusion, and stagnation since the 2000s. Although this study does not cover the twenty-first century, chapter 7 discusses the inception of this confusion. Nonetheless, Downey et al. (2007, 474–481) broadly considered Japanese engineering ethics since the late nineteenth century by analyzing ethics as obligation and responsibility for national and corporate households (familism, "ie" in Japanese) and their impasse. This notion is common in Japanese organizational culture. However, such an analysis tends to equate engineering ethics with business ethics and does not fully reveal the particularities in Japan. In the 1990s, engineering ethics was expected to be a new type of Western professional ethics, distinct from the business ethics of traditional Japanese culture. Thus, to understand Japanese engineering ethics, certain expectations must be analyzed in-depth. However, other historical studies have only outlined these historical processes in the United States and Japan.[7] While many essays have presented author experiences, most studies have not examined the original historical materials.

Hence, this study provides a consistent historical picture of the introduction of engineering ethics in Japan by comprehensively discussing its development throughout the 1990s and the entire twentieth century.[8] This analysis emphasized the social contexts that have promoted engineering ethics rather than the specific content required by the promoters for each period.

In particular, this study focuses on the relationship between engineering ethics and democracy. Democracy became a basic principle of the social reforms in postwar Japan, following its suppression in prewar Japan. The Gijutsushi system and its code of ethics were established to advocate for democracy, given Japan's defeat in the war. On the first page of its inaugural journal in 1951, the JCEA declared:

> Democracy has begun in the new Japan. Industrial reconstruction will depend on the breakthrough in feudal technology. For equal access to superior technology for all companies, the flag of the democratic movement for technology has been raised. The pioneer of the technology revolution [. . .] is Gijutsushi. (JCEA 1951a, 1)

While engineers' professionalism does not generally concern democracy, it fits within democratic ends, because it emphasizes the ideal of an autonomous individual, which is essential to democracy (Layton 1971, 7). Nevertheless, the elitist professionalism of engineers included notions of social hierarchy and inequality, whether in membership grades or society in general (Layton 1971, 65). In this dilemma, the Japanese Gijutsushi system preserved the exclusiveness and peculiarity of the organization and entered the 1990s without being popularized. This study provides a historical understanding of the efforts of Japanese engineers to improve their social status within the context of engineering ethics.

Moreover, since the economic-globalization-inspired educational reform introduced engineering ethics, considering the Japanese problem of industry-academia cooperation is important. Japan's engineering education reforms promoted cooperation with the industry; however, a long-term alienation existed between the two in postwar Japan. Inose Hiroshi, an engineering professor at the University of Tokyo, reported at the United States–Japan conference in 1981[9] that industry-academia cooperation had been "taboo" in Japan since the 1960s; a strong antiwar sentiment against the prewar promotion of industry-academia cooperation erupted with the 1960s student activism. Furthermore, the industry-academic difference in research values sowed distrust therein (Inose et al. 1982),[10] which seems to have variously influenced engineering education reform in the late twentieth century.

In summary, this study discusses the relationship between engineering ethics and social and cultural perspectives throughout the twentieth century, such as democracy issues, economic globalization, engineers' social status, and industry-academia cooperation. These topics have not been adequately discussed in previous studies. Chapter 1 discusses the situation in Japan regarding ethics for engineers before World War II, especially the case of the JCES. Chapters 2 and 3 discuss the engineering ethics introduced via the Allied Occupation regarding engineering education and qualifications, respectively. Chapter 4 examines the development and criticism of industry-academia cooperation from various social perspectives. Chapter 5 describes how the cooperation led to engineering education reform in the late 1980s. It also discusses the development of applied ethics and the process of establishing the IPSJ's code of ethics. Finally, chapters 6 and 7 analyze the process of introducing engineering ethics through the globalization of engineering qualification and education.

NOTES

1. The 7.3-magnitude earthquake that struck the Kobe area on January 17, 1995, caused tremendous damage with 6,434 deaths, 3 missing persons, and 43,792 injured. It caused massive fires and the collapse of buildings, including highway bridges. Just a year before that, on January 17, 1994, the Northridge 6.7-magnitude earthquake had occurred in the United States, also causing the collapse of highway bridges; however, Japanese civil engineers had announced that such a collapse would never happen in Japan.

2. During the morning commute on March 20, 1995, devotees of the Aum Shinrikyo doomsday cult released sarin gas, a chemical weapon, on Tokyo subway trains, causing 13 deaths and leaving more than 5,800 people injured. In the previous year, they released sarin gas in a residential area in Matsumoto City, Nagano Prefecture, causing 8 deaths and leaving 140 injured. It was a great shock to Japanese society that people with higher education in science and engineering, who were supposed to be rational, joined such a cult and committed acts of terrorism.

3. On December 8, 1995, the Monju fast breeder reactor was shut down due to the leakage of sodium coolant. The Power Reactor and Nuclear Fuel Development Corporation (PNC), the operator, concealed the sodium leak section in the footage of the accident, for fear of increasing public criticism, and the head office of the PNC was involved in the cover-up. Moreover, a staff member who conducted the internal investigation committed suicide.

4. On September 30, 1999, a criticality accident occurred during an enriched uranium processing at JCO's nuclear fuel-processing plant in Tokaimura village, Ibaraki Prefecture. Three workers were exposed to high levels of radiation, two of whom died. The site workers tampered with the official manual to prepare a secret manual and used a stainless-steel bucket to illegally process the uranium solution. The evidence stunned the public.

5. An analysis of newspaper articles since the 2000s illustrates that few writers or journalists discuss engineering ethics, especially from 2008 to 2016, thus representing a decline in public interest since the late 2000s (Yamaguchi and Shibata 2017).

6. Apart from professionalism, the National Labor Relations Act (the Wagner Act) was enacted in 1935 to protect the rights of employees and employers. It encouraged collective bargaining.

7. For example, see Kikuchi (2004) and Fujiki (2011, 2012).

8. This study limited the engineering education analysis to higher education. Technical schools also provided ethics education or moral training; however, it was a part of the moral education required in general middle-level education. Including them in this study necessitates a discussion of the entire history of moral education in Japan. However, as technical schools educate technicians and technologists rather than engineers, the study will deviate from the subject of global standards. Therefore, this study focuses on college-level engineering education.

9. This is the second seminar on science policy under the U.S.-Japan Cooperative Science Program held in Hawaii.

10. Considering the industry-academia cooperation as research commissioned to universities, such phenomena became global when the usefulness of science to the industry became apparent. Moreover, in Japan, the establishment of affiliated research institutes, scholarships, and commissioned research funded by corporate donations became popular around the time of World War I. Prior to this period, the Furukawa Zaibatsu conglomerate donated the construction budget to the Imperial Universities of Kyushu and Tohoku. Afterward, in 1930, the Resources Council recommended the promotion of cooperative research by establishing a research union of researchers and entrepreneurs. Moreover, in 1940, the system of neighborhood research groups was introduced. However, the former was the return of corporate profits to the state, and the latter was a part of the national mobilization system of science and technology. Neither was the kind of investment partnership expected today (Kitami 1962, 9; Tachi 1983, 8–9). After the war, companies introduced technology from the United States and European countries without relying on universities. Furthermore, they promoted their research and development (R&D) systems by establishing research institutes. With the expansion of science and technology promotion in the late 1950s, industry-academia cooperation gained social attention, and the support from the industry for engineering faculties increased. However, toward the end of the 1960s, the criticism of industry-academia cooperation was widespread and conflicted with academic independence (Hashimoto 1999; Li 2012). Chapter 4 discusses this in more detail.

Chapter 1

Engineering Ethics in Prewar Japan

JAPANESE ENGINEERING ETHICS BEFORE DEMOCRACY

Before World War II, the extent of emphasizing moral education within Japan's engineering education was identical to that of its general middle-level education.[1] Professional ethics for engineers were introduced in engineering schools; moreover, the Japan Civil Engineering Society (JCES) established a code of ethics modeled after the American Society of Civil Engineers (ASCE). This chapter discusses Japanese engineering ethics from the Meiji era (1868–1912; when Japan lifted its isolation policy on trade with Western countries and promoted modernization) to the end of World War II and from the engineering education and social status of engineers' perspective.

The industrial promotion policy of the Meiji Restoration required the training of engineers who were familiar with Western science and technology.[2] The Ministry of Industry (Kōbushō) established the Engineering Board (Kōgakuryō) in 1871 and reorganized it as the Imperial College of Engineering (Kōbu Daigakkō) in 1877. As Japan aimed to import Western science and technology, engineering education in the early Meiji period was entrusted to *oyatoi gaikokujin* (employed foreigners) engineers. In particular, Henry Dyer, who was only 24 years old and studied under William J. M. Rankine at the University of Glasgow in Scotland, came to Japan as the principal of the Imperial College of Engineering and designed its unique curriculum. Dyer's pedagogy was a hands-on learning program (later called a "sandwich" system), which combined lectures and experiments on campus with on-site activities as part of the Ministry of Industry projects, modeled on British practical and French theoretical styles of engineering education. However, continuous and consistent science education became insufficient within the practical education program (Kakihara 1996, 3–9). Thus, the Imperial College began to require a graduation thesis from 1883 (the year after Dyer returned to Scotland) and began transitioning from a vocational school to an academic college. Meanwhile, the Ministry of Education (MOE) set up an engineering department at the University of Tokyo's Faculty of Science in 1877, emphasizing design education in conjunction with basic subjects, such as physics. However, after the department was upgraded to a faculty in 1885 and integrated with the Imperial College of Engineering to form the College of Engineering of the newly established Imperial University (given the abolition of the Ministry of Industry), the curriculum shifted from practical activities and focused on academic lectures and exercises (Oyodo 2009, chap. 1; Miyoshi 2005, 216–218).

While universities promoted academic programs, technical schools encouraged practical ones. Tokyo Technical School (currently Tokyo Institute of Technology) conducted a great deal of on-campus factory training since its establishment. Furthermore, Osaka Higher Technical School (currently the Faculty of Engineering at Osaka University) designed a "sandwich" program of practical training in 1905 in the first semester of the first grade and the second and third semesters of the final grade (Osaka Higher Technical School 1905a, 32; 1905b, 30). Meanwhile, in the United States, the University of Cincinnati reported industry-academia cooperative education (a sandwich program) in 1906.[3] However, the curricula of Japanese technical schools did not provide on-site practical training in factories via cooperation with private companies. Instead, they provided on-campus training. The industry cooperative education program was merely a factory tour for students during their vacations.

In the early Meiji period, the Japanese government constructed a new education system to introduce Western knowledge, but conservatives claimed the bias toward knowledge-based education caused moral confusion in society. In response, Emperor Meiji published a traditional and Confucian-based declaration, the Imperial Will on the Great Principles of Education (Kyōgaku Seishi), in 1879, which demanded that *jingi chūkō* (benevolence, righteousness, loyalty, and filial piety) should form the core value of education rather than *chishiki saigei* (science and technology). He also promulgated the Imperial Rescript on Education (Kyōiku Chokugo) to emphasize moral education in 1890. Inoue Kowashi, director-general of the Legislation Bureau, drafted the Imperial Rescript based on the *Kokutai* (Japan's national polity) pedagogical policy, arranged by modern legal studies following the German monarchy. It rejected Rousseau-like popular sovereignty of the freedom and people's rights movement (*jiyū minken undō*) and reformed the premodern Confucian virtue-oriented loyalty education of the Imperial Will.[4] *Kokutai* in modern Japan, like Christianity in Western countries, was the constitutional idea of establishing Japan as a constitutional monarchy by granting sovereignty only to emperors of an age-old unbroken line under the State Shinto religion. The *Kokutai* pedagogical policy was centered on moral education and valued traditional and spiritual integration to promote loyalty and patriotism to Japan as a nation-state.

Although no ethics class was provided at universities specializing in respective academic fields, it was introduced into vocational education at technical schools. Inoue Kowashi, the MOE minister, emphasized the importance of vocational education, especially engineering education, for the nation's wealth and called for moral education to contribute toward national interest by eliminating the old culture of merchants (Noguchi 1994, 447–485).[5] He delivered the following speech at the graduation ceremony hosted by Tokyo Technical School in July 1894:

> Vocational education should be based on moral education . . . the person who tries to promote our still infantile businesses for public interest and advance the source of the nation's economy, regardless of the size of the business, must be grounded in honesty, clean, and trusted by society . . . if a corrupt who lacks even one of these precious virtues were to be allowed to invade the society of industry and technology . . . vocational education would soon lose its momentum and progressing businesses would ultimately collapse and fall into a helpless situation. (Tokyo Institute of Technology 1940, 191–192)

As described in the introductory chapter, against the backdrop of his appeal that "vocational education should be based on moral education" was the devastation of vocational ethics in the industry, which became a social issue, and the cause was allegedly the collapse of the premodern apprenticeship system. As the system was overseen by the Ministry of Commerce and Industries, the MOE focused on the cultural aspects of vocational education and attempted to rebuild vocational ethics (National Institute for Educational Research 1973, 227–229).

Per the policy of emphasizing moral education, Tokyo Technical School began a one-hour ethics class held every other week in all semesters in 1899. The school explained the contents of the class as follows:

> The main purpose of ethics is to make the moral judgment of engineers clear and reliable for practical application. That is, the course refines a student's moral common sense about appropriate practical problems in daily life and develops it into a certain organizational knowledge to be finally applied to special issues in the industry to clarify the industry morals. (Tokyo Technical School 1899, 11)

Although the details are not clear, this explanation demonstrates that the school taught professional and practical engineering ethics.

In 1895 and 1905, Japan won the Sino-Japanese and Russo-Japanese Wars, respectively. The victory over Russia, a major Western power, and China, traditionally an advanced superpower relative to Japan, greatly enhanced their pride in the international community. The Bushido (the Samurai Way) theory flourished simultaneously to demonstrate the superiority of the traditional morality of Japan relative to Western countries through the reinterpretation of history, thereby building Japan's national ego. In particular, Nitobe Inazo's book, *Bushido: The Soul of Japan*, published in English in 1900, translated Western ethics into Japanese warrior virtues. The historical interpretation was borderline fiction, because he did not adequately examine the historical details of real samurai values. However, because the book was based on Western ethics, it was featured in various Western media, and Bushido became famous. Bushido was framed regarding its unique spirituality against the materialistic and mechanistic power of Western science and technology but was compatible with the Western Stoic thought, chivalry, and gentlemanliness (especially British gentlemanship). The fact that Japan defeated a Western power in the Russo-Japanese War in 1904–1905 despite its inferiority to Western countries in science and technology, seemed to prove the alternative non-scientific explanatory principle. Therefore, the Japanese adhered to Bushido as an idea to supplement moral education that was expected to be lacking along with modernization via Western science and technology.

The revival of moral education was a worldwide trend during this period. The International Union of Ethical Societies, established in 1896 against the decline of religious authority, held the First International Moral Education Congress in London in 1908, where 192 organizations, mainly from Europe and the United States, participated. Hojo Tokiyuki (1913) of Japan reported that Japan's moral education was based on loyalty and filial piety. He learned about the British Boy Scout movement, which had just begun the previous year, and introduced it to Japan as "Shōnen Musha" (Boy Samurai). The Boy Scout movement gained popularity in Japan as an activity with ideas similar to Bushido.

Meanwhile, the expansion of individualism and socialism became prominent with modernization in Japan, and reports of corruption of students' moral discipline were frequent. In 1908, Boshin Shōsho (Imperial Rescript for Japanese discipline) was promulgated to address the situation as a social disorder in the post–Russo-Japanese War period and remind the population to respect the Imperial Family, work dutifully and single-mindedly to develop the nation's destiny, and strive for diligence and frugality. Komatsubara Eitaro, who became the MOE minister in the same year, promoted more substantial moral education for secondary education levels. In 1910, the second edition of a national textbook for moral training was published per the extension of compulsory education because of the rise in the school enrollment rate.[6] The strengthening of nationalist moral education became an urgent issue due to the High Treason Incident (Taigyaku Jiken) that occurred in the same year, where many socialists and anarchists were arrested on suspicion of plotting to assassinate the emperor.

In response to the social demand, a general nationalist moral education under the *Kokutai* pedagogical policy became necessary at technical schools as well as ordinary schools. Osaka Higher Technical School also began to hold a one-hour ethics class every other week in 1909. Furthermore, both the Tokyo and Osaka schools increased the frequency of ethics classes to once every week and changed the subject name from ethics (*rinri*) to moral training (*shūshin*) in 1911 according to the nationwide enhancement of moral training. The Osaka class taught the principles and practical morals needed to cultivate sociability and individual autonomy, value public over private interest, respect the Imperial Family and the national constitution with loyalty and filial piety, and develop the family business as an engineer to build national wealth (Osaka Higher Technical School 1911, 45).

The Special Council for Education (Rinji Kyōiku Kaigi), established in 1917 as an investigative and deliberation committee directly under the Cabinet of Japan, recommended that national moral education should be based on the national polity and strictly implemented to cultivate character-building, thereby avoiding bias toward skill-based education and promoting moral training for vocational schools. The University Order, enacted in 1918 to reflect the council's report, clearly introduced the *cultivation of nationalism* as its major modification. The revised Vocational School Order of 1920 also mentioned the cultivation of moral character.

Specialized academic research was increasingly emphasized to improve Japan's industry and its independence since World War I. When higher technical schools were upgraded to universities in 1929 to expand universities, moral training classes were abolished, and third-year practical training was replaced with graduation theses. Moral training at higher technical schools was inadequate to the institutional requirement for vocational education. Given that higher education aimed, as specified in the University Order, to teach the academic theories and applications (that were indispensable to the state) and research the esoteric, engineering ethics classes were never offered at university.

In the same period after World War I, Hirayama Fukujiro (1926), a civil engineer who led the Japan Consulting Engineers Association (JCEA) after World War II, proposed to promote "conceptual education" in other fields, such as law, economics, sociology, and philosophy and organize social activities, such as student clubs, because engineering education disregarded humanities education relative to elementary and middle schools (Hirayama 1926). Yagi Hidetsugu (1926), one of Japan's leading electrical engineers, also stated that ethics education was the most critical deficiency of the country, and character-building was the most important element in engineering education. The curriculum of technical schools saw an expansion in ethics education, but the insufficiency remained a subject of discussion. In particular, the policy of Japanese university education, which emphasized academic rather than vocational education, was disconnected from the students' goal of gaining employment. This difference continued to be a critical issue in the democratic engineering education reform after World War II. Chapter 2 discusses the post–World War II engineering education reform in detail.

YAMAKAWA KENJIRO'S BUSHIDO-GENTLEMAN MORAL EDUCATION

Yamakawa Kenjiro, who served twice as the president of Tokyo Imperial University (1901–1905 and 1913–1920), was a leading figure in promoting moral education in engineering education from the Meiji through to the Taisho era (1912–1926). He became the first Japanese professor of physics in Japan after its introduction as an academic subject and served as the president of the Imperial Universities of Tokyo, Kyushu (1911–1913), and Kyoto (1914–1915). He was also the first president of Meiji College of Technology (MCT, Meiji Senmon Gakkō, currently Kyushu Institute of Technology), a private higher technical school, established in 1907. As the MCT president, he established a policy to emphasize nationalist moral education, *shikunshi* (gentleman) education, derived from British gentlemanship and Japanese Bushido. He was also the head of the Investigation Committee's first division, which was in charge of the second edition of the national textbook for moral training in 1910, and a member of the Special Council for Education in 1917.[7]

The civil wars in Japan and the United States influenced Yamakawa's sociological perspective. The Boshin War (1868–1869), where the new government forces subdued old shogunate forces immediately after Japan entered the Meiji era, and the American Civil War (1861–1865), where the North Union and the South Confederacy fought over the abolition of slavery, were particularly influential.

Born a samurai of the Aizu clan (currently in Fukushima Prefecture) at the end of the Edo period, Yamakawa experienced the Boshin War as a member of the Aizu Byakko Tai (White Tiger Unit), a young samurai unit under the age of 18, on the side of the old shogunate forces. Defeat in the war drove the Aizu samurais out of their country to the northern and extremely cold Tonan clan (currently in Aomori Prefecture). Subsequently, he went to the

United States to study on a government scholarship in 1871 at the age of 18 and entered the Sheffield Scientific School at Yale University in 1872. After completing his degree, he returned to Japan in 1875 and became an assistant professor of physics at Tokyo Kaisei School (Tokyo Kaisei Gakko) and the University of Tokyo. In 1879, at the age of 26, he became a professor of physics at the University of Tokyo. In his later years, he explained why he decided to study physics in the United States:

> According to Youmans' thought, I have to improve politics to make Japan flourish by all means. To improve politics, society must be improved. To improve society, it is necessary to study sociology. The study of sociology requires the study of biology and other natural sciences. So, after all, I decided to study physics because I thought that physics and chemistry must flourish in order to enrich our country and strengthen the soldiers. (Yamakawa [1929] 1937d, 53)

Yamakawa stated that he first had a national purpose and chose physics according to the hierarchical classification of academic fields: physics-biology-sociology.[8] Edward L. Youmans, an American writer and editor, launched the *Popular Science Monthly* journal in 1872 and published top priority papers titled "The Study of Sociology" by British philosopher Herbert Spencer, thereby promoting his ideas. In 1929, Yamakawa ([1929] 1937d, 53) remembered: "I have never thought about my future as much as that time. A book on a new philosophy of a man named Herbert Spencer came out and had a great influence on the thoughts of young people." Spencer introduced the term "survival of the fittest" to explain the evolution of organisms in terms of scientific law and developed his sociological theory to discuss the evolution of society, which he considered an organism, thus popularizing the term "evolution" (Spencer 1864–1867). Youmans vigorously introduced Spencer's sociology in the United States, where he was described as an "apostle of evolution" (Fiske 1894, chap. 9).

The experience of studying abroad strengthened Yamakawa's nationalism. Despite the Emancipation Proclamation ending slavery during the American Civil War in 1863, the Ku Klux Klan's (KKK's) lynching of African Americans and many other horrendous acts occurred. He got the impression that the United States was "a very outrageous and barbarous country" because the KKK was not prohibited by law (Yamakawa [1929] 1937d, 70–72). The KKK was subject to punishment under the federal law in 1870–1871 when Yamakawa visited the United States, but the lynching continued. He was strongly aware of the importance of the nation-state for the protection of the people.

Thirty years later, in 1901, Yamakawa became the president of Tokyo Imperial University when Bushido began to attract social attention. Having assumed the university president role, he began to develop his educational philosophy modeled on Bushido, especially during the Russo-Japanese War. He translated British "gentleman" as "shikunshi" and presented gentlemanship and Bushido as similar moral codes (Yamakawa [1905] 1937a, 191). Yamakawa listed eight common norms between the two: (1) do not lie and be responsible for your remarks; (2) respect for the sake of honor; (3) do not act meanly; (4) do not change your behavior frivolously according to your interest and keep your integrity; (5) respect for the sake of faith, friendship, loyalty, and fidelity; (6) engage in fair play; (7) respect for the sake of courage; and (8) comply with courtesy. Nevertheless, he also noted the difference as follows: (1) respect for women in gentlemanship relative to (2) loyalty and filial piety to the monarch and parents with preparedness, as well as respect for death, in Bushido (Yamakawa [1905] 1937a, 191–192).[9] *Chūkō* (loyalty and filial piety) was considered a traditional and essential moral norm in support of Japan's national polity, as found in Bushido.

MCT, which opened in 1909, respectfully adopted Yamakawa's ethics. He was entrusted with policymaking and assumed the presidency to oversee the entire education system when the school was established. He introduced his unique moral education policy and declared that MCT was "a school to produce not merely technicians but also *shikunshi* who are acquainted with technology" (Yamakawa 1911, 52). He also developed a four-year boarding school education system that emphasized science (as the basis of engineering), English, and military training (gymnastics). His emphasis on military training, and moral education came from his experience of a bitter victory in the Russo-Japanese War. Despite suffering massive casualties, heavy tax increases, massive debts, and ultimately winning the war, Japan could not obtain monetary compensation under the Treaty of Portsmouth, which increased dissatisfaction in Japan more than that during the Triple Intervention by Russia, France, and Germany after the Sino-Japanese War in 1895 (a demand to abandon the Liaodong peninsula, one of the spoils of the war). This dissatisfaction exploded across the country, and campaigns against the peace treaty broke out, such as the Hibiya incendiary incident[10] in 1905. At Tokyo Imperial University, Tomizu Hirondo and six professors gave radical speeches to improve the conditions of the peace treaty or continue the war. The Cabinet feared the campaign's social impact and suspended Tomizu, causing protests against power abuses within the university. Taking responsibility for the university autonomy disturbance, the MOE minister and President Yamakawa resigned (known as the Seven Professors Incident). As president, Yamakawa was in the position to reprove Tomizu and others; hence, his resignation. However, he likely shared the social dissatisfaction at the time.

After the Russo-Japanese War in 1905, international relations entered a new phase. Yamakawa strongly feared the growing conflict between the

Triple Alliance (signed by Germany, Austria-Hungary, and Italy in 1882) and the Triple Entente (developed by Great Britain, France, and Russia in 1907). At the opening ceremony of MCT, he listed five conditions of a strong country in the age of the survival of the fittest: (1) a large country, (2) a large population, (3) a rich country, (4) a knowledgeable populace, and (5) an intensely patriotic populace. He claimed that "the only weapon our country can use is the patriotism of the people" because Japan ranked at or was near the bottom of the first four conditions relative to the powers of the countries of the Triple Alliance, the Triple Entente, and the United States (Yamakawa 1911, 53). He appealed to patriotism as Japan's ultimate weapon:

> Both the Sino- and Russo-Japanese Wars proved the strong patriotism of the Japanese. In war, there is no single opponent. It is the so-called national unity. Those staying in the country are willing to carry the burden of almost unbearable taxes. Those in the battle treat their lives as light as dust and confront the enemies. It is not inferior to the people of other countries, and they admire it. (Yamakawa 1911, 53)

Yamakawa requested that students be patriotic to compensate for the weakness of the Japanese military potential relative to other countries. He maintained that "If we the people lose our patriotism, we will lose the only weapon that keeps us alive; so we must prepare ourselves for the time when our ethnic nation-state will vanish" (Yamakawa 1911, 54). His understanding was later reflected in the dormitory names of MCT, which came from the old Chinese proverb "kokuji, bōka, kōji, bōshi" (forget your family for the state and forget your private life for the public).

This sense of crisis regarding international relations was supported by the theory of *survival of the fittest*, both biologically and sociologically. Spencer's social philosophy affected a wide range of political positions in Japan, from the Meiji government to the freedom and people's rights movement. One of those influenced was Kato Hiroyuki, who served as the president of Tokyo Imperial University during the 1877–1886 and 1890–1893 periods, lectured at the university and developed a corresponding sociological term "yūshō reppai" (the superiors win and the inferiors are defeated). Others included Toyama Masakazu, who served as the president during the 1897–1898 period, and Ernest F. Fenollosa, an American historian of oriental art. Yamakawa emphasized that "yūshō reppai" was a scientific principle in his presidential address at Kyushu Imperial University in 1912:[11]

> Needless to say, the fittest survives and the unfit dies, *yūshō reppai*, in short, is the main principle of biology. All living things, whether animals or plants, are

under this principle; so, this principle applies not only to human beings but also to the nation-state as a human organization. (Yamakawa [1912] 1937b, 242)

Spencer's sociology, reinforced by issues of the global escalation of racism and a strong sense of caution against invasion by other countries, profoundly impacted Japan. The units of invasion and defense were ethnicity, race, and the nation-state. These reasons were quite different from that of the American engineers' case, as described in the introductory chapter. At the graduation ceremony of Kyushu Imperial University the following year, Yamakawa explained the need to protect the nation-state by considering the example of the Jewish persecution in Russia (namely Pogrom):

> No one in the world is as unhappy as people without their country. Although "without their country" does not mean "without their nationality," they are people who are treated as foreigners or less than foreigners by other people in the country where they live, consider themselves to be spongers, and do not have the same rights as other citizens. The awful examples are black people in the U.S. and the Jewish people in Russia. (Yamakawa [1913] 1937c, 245)

Yamakawa harbored the impression that the United States was "a very outrageous and barbarous country," given his experiences while studying there. In Russia, as with many pogroms, the claim of "yellow peril" increased. He believed that Jewish people escaped extinction because of their academic excellence. He felt a sense of crisis, fearing that Western countries could persecute the Japanese because they were academically inferior relative to the literary works by the Jewish people. He continued: "If there is anyone to conquer our country by any possibility, it is none other than white people who believe in Christianity. From the perspective of white people, it is clear that we are a different race, pagan, and even more detestable than the Jewish people" (Yamakawa [1913] 1937c, 251).

In 1917, the Special Council for Education was established in Japan, and the Russian Revolution broke out, followed by the collapse of the Russian Empire and the establishment of the Soviet Union. The expansion of communism enhanced Yamakawa's sense of crisis further. Although socialism spread in Japan, he rejected it. He was forced to respond to the Morito Incident[12] in 1920 during his second presidency at Tokyo Imperial University. Yamakawa stood by Morito's academic autonomy, as well as during the 1905 Tomizu Incident. However, the social situation worsened, where merely writing a paper on anarchical communism in an academic journal was sufficient for an author to be imprisoned for rebellion against the government. Yamakawa expressed the following negative opinion in his retrospective reflection of his time as a student in the United States:

Despite the huge difference in wealth between America and Japan, I had concluded that Japan would win against any other country in the world, after all. There were Socialist Parties in foreign countries at that time. They were expanding in power day by day. The governments would be destroyed by Socialist Parties, democracy, and Communist Parties. When those happened, the countries weakened. I believed that the Communist Party would never be formed in Japan given its three-thousand-year history. This kind of government must have been weak because foreign countries could not build strong governments by democracy or Communist Parties. I thought we could win after all because Japan was different. I believed in this from that time until very recently; but unfortunately, the Communist Party movement has happened in Japan, and my belief was betrayed. Students studying abroad at that time seriously believed that they had to do their best for the country. (Yamakawa [1929] 1937d, 55)

Yamakawa advocated for academic autonomy but was strongly concerned about the expansion of communism, especially after the incident on March 15, 1928, when members of the Japanese Communist Party were arrested across the country. At this time, he took the initiative to counter the spread of democracy and the communist movement through policymaking in education.

Yamakawa was among the leaders engaged in strengthening moral education based on the *Kokutai* pedagogical policy. His argument was based on general nationalist morality, and he never mentioned the professional ethics of engineers as far as his records show. He promoted the strengthening of moral education as a countermeasure to modern social ideas that were popular at the time, such as democracy and communism. His moral education for engineering students was contrary to individualistic ethics. In contrast, however, the spread of democracy led engineers to organize a new social movement, which drew their attention to American engineering ethics. The next section discusses the relationship between democracy and the movement of the engineers.

TAISHO DEMOCRACY AND THE ENGINEERS' MOVEMENTS

Just before the Taisho era, as modernization progressed, Japanese citizens increased their awareness of rights, and a social trend called "Taisho Democracy" was born. The corresponding idea was *minpon shugi* (democracy without popular sovereignty), which was proposed by Yoshino Sakuzo in 1916. The Japanese government considered popular sovereignty as a dangerous and unacceptable idea because it could subvert the national polity of the emperor's sovereignty. Yoshino, thus, assumed that democracy had two meanings—he interpreted "minpon shugi" as that "the basic goal of national sovereignty activities should be for the people from a political perspective" and distinguished it from radical "minshu shugi" that shifted the subject of discussion to sovereignty itself, as "the sovereignty of the state should be in the people from a legal perspective" (Yoshino 1916, 39).

Nitobe Inazo, who published *Bushido*, also proposed in 1919 that the word "democracy" should be translated as "Heimindo" (the Citizens Way). Although he set up Heimindo as an extension of Bushido, he saw Bushido as obsolete. He eliminated social class distinctions and argued that a moral code that considered peace as the ideal and normal state, rather than armed force, was appropriate for the morals of the general public per the principles of the League of Nations (Nitobe 1919, 18). The trend of Taisho Democracy also activated civil movements. In 1918, the speculative heavy buying of rice in anticipation of demand in the Japanese Siberian Intervention became rampant, and the price of rice soared, causing the Rice Riots (Kome Sōdō) to spread across the country. The riots and the 1917 Russian Revolution spurred the expansion of Japan's social movements.

The increased awareness of rights improved working conditions. In Japan, the 1911 Factory Act, which came into force in 1916, restricted working hours, prohibited child employment and late-night work for the young and women, and stipulated employers' obligation to provide accident compensations to workers. The International Labour Organization (ILO), established under the Treaty of Versailles in 1919, began to improve working conditions worldwide. In Japan, the rapid industrial development during World War I led to the rise of the working class, the expansion of labor movements, and the rapid increase in labor disputes.[13]

Workers' protection in Japan was facilitated under the principles of *onjō shugi* (paternalism) and humanitarianism at that time. *Onjō shugi* was a management-familism that regarded the relationship between employer and employee as a parent-child relationship sharing a common interest, where cooperation was not based on labor-management equality but emphasized harmony based on hierarchical relationships (Hazama 1978, 9–10). The patriarchal system that extended nationwide from the emperor to each family encompassed everyone and was considered a national polity in Japan. The Japanese civil code was also based on this idea. In 1919, the Cooperation Society (Kyōchōkai) was founded by Tokugawa Iesato, head of the former Tokugawa Shogunate family, and Shibusawa Eiichi, a business leader. It promoted the labor-management unification movement based on *onjō shugi*. Meanwhile, the Japan Federation of Labour (Nihon Rōdō Sōdōmei) held the first May Day in 1920 and encouraged a class struggle.

A typical example of the management-familism is the case of the Japanese government railways. Uchida Nobuya (1935, 7–8), the minister of railways, reduced the concept of "railway spirit" (tetsudō seishin) to the traditional *Japan spirit* (Nihon seishin). The Japan spirit or Japanism (Nihon shugi) was a nationwide concept of familism with the emperor and empress as parents and the subjects (citizens) as children. Although his railway spirit was ambiguous, according to an analysis of the code of practice for the railway staff adopted in 1925, the values in all twenty-eight clauses were reduced to three: *kenshin hōkō* (devotional service), *wagō keiai* (respectful harmony), and *shūyō renma* (disciplinary training). They were summarized into a single concept: large-familism (Oshima 1935, 194–200). The code communicated an understanding that employees should be content with their lot, and it was desirable to serve the favors gifted from the public and compromise with others without resorting to self-assertion, while also respecting the existing order in relationships (Hazama 1963, 135–141). The Japanese government railways conducted education and communication activities on various occasions to embody the *onjō shugi* and familism as ethical standards for employees.

However, along with the rise of the labor movement, several movements launched to improve the social status of engineers.[14] Naoki Rintaro, who became a director of the harbor facilities section in Tokyo after working as an engineer for the Ministry of Home Affairs,[15] contributed many articles on social issues regarding engineers to the *Kōgaku* (*Engineering*) magazine, which was first published in 1914. In 1918, he arranged and published these articles together as one book titled *From Technology Life* (Gijutsu seikatsu yori). He argued that the social status of the brand-new *engineer* class was not properly recognized in Japan, because engineering was the newest professional field in the world and engineers had no awareness of their social roles in daily activities (Naoki 1918, 2–3). He called for the will and cooperation of engineers as the new knowledge-based class with examples from other countries, such as the Association of German Engineers (Verein Deutscher Ingenieur), the ASCE, and the AICE.

In Japanese civil service, the Civil Service Appointment Ordinance (Bunkan Ninyō Rei) was enforced in 1893, which, as a general rule, produced bureaucrats to be selected in the Civil Service Higher Examination (Bunkan Kōtō Shiken).[16] The ordinance aimed to prevent people without appropriate expertise or experience from becoming bureaucrats via political power. On the one hand, the direct access law graduates, from imperial universities, possessed to become bureaucrats without examination was abolished. On the other hand, appointing law graduates to important posts in bureaucratic personnel affairs had become stronger, given that the modern administration should be organized via legislation. However, the social importance of engineering had also increased since the twentieth century; electrical, mechanical, and applied chemical engineers were more valued in the industry. Accordingly, civil engineering technocrats, such as Naoki, became aware of the lowliness of their social status. At the time, engineer positions of the director-general and above were restricted to the chief engineer of the Ministry of Home Affairs and the director-generals of the Ministry of Railways, the Engineering Bureau of the Ministry of Posts and Telecommunications (MOPT), and the Construction Bureau of the Navy Ministry. Not even the director of the Civil Engineering Bureau of the Ministry of Home Affairs was served by an engineering official; he was served by an administrative official. The chief engineer was supposed to enjoy the same level as that of the deputy minister in the Ministry of Home Affairs. In reality, however, he was treated like a subordinate of the director-general of the Civil Engineering Bureau, which enforced the criticism that engineers were discriminated against for raises in salary and authority and that they were devalued as mere craftsmen.

The Engineering Society (currently the JFES) submitted a proposal to the prime minister to amend the Civil Service Appointment Ordinance, asking for access to higher posts for technocrats. The government's answer, however, only followed the conventional policy of limiting the work of technocrats to engineering expertise. Okochi Masatoshi and Shiba Chuzaburo, who were engineering professors at Tokyo Imperial University and members of the House of Peers, then established the Association for the Promotion of Industrial Problems (Kōseikai) in 1918 to promote the union and independence of the industry and develop activities to improve the industry and status of engineers. The agenda included the promotion of engineering education and its institutional reform.[17]

Okochi was appointed as the director of the Institute of Physical and Chemical Research (RIKEN) in 1921. RIKEN was established in 1917 as a national basic research institute to develop the Japanese heavy chemical industry and procure industrial goods domestically, which became an issue during World War I.[18] RIKEN aimed to "conduct original research on physics and chemistry to contribute to the development of industry and apply its achievements" (RIKEN 1917); therefore, it also aimed to acquire patents and utility models by promoting applied research. However, it was unable to raise the expected amount of donations from the industry and, thus, fell into financial difficulty. Amid this challenge, Okochi took over as the director and proceeded with organizational reforms. He founded the Industrial Company for RIKEN (Rikagaku Kōgyō) to commercialize and industrialize inventions and patents in 1927, expanding it to the RIKEN Group (RIKEN Konzern), which comprised various related companies.

Okochi was critical of the fact that scientific investment had been postponed in the old-fashioned capitalist industry characteristic of Britain, because managers without scientific knowledge colluded with veteran mechanics and attempted to do well only with passive adjustments of production volume, prices, and wages. His argument was an indirect criticism of Japan's traditional *zaibatsu* conglomerate (large business conglomerate in prewar Japan) and an appeal to the new RIKEN Konzern system. He advocated for a *kagaku shugi kōgyō* (science-based industry) or *chinō shugi kōgyō* (knowledge-based industry), contrary to the traditional style. He maintained that it was possible to develop businesses, like the nitrogen chemical industry, systematically, even in a country without resources (Okochi 1938, 3–77). Although his science-based industry differed on whether it hinged on a scientist or an engineer, it was similar to the *technocracy* argument coined by William H. Smyth in 1919 and was systematically developed by Thorstein B. Veblen in 1921.[19] Thus, from the 1910s to 1920s, mainly in the Taisho era, Japanese engineers began to develop a movement to emphasize science and engineering and improve their social status.

Through the spread of Taisho Democracy, Miyamoto Takenosuke, a civil engineer and technocrat, organized the Japanese Technicians' Union (Nihon Kōjin Club) in 1920 with the alumni of the Department of Civil Engineering at Tokyo Imperial University and engineers of the Civil Engineering Bureau of the Ministry of Home Affairs and launched a movement to improve the social status of engineers against the capitalist industry.[20] The initial number of union members was about 200 but grew to 3,500 in one year and reached 5,500 by the beginning of the Showa era in 1926. He stated that the organization's "leading spirit" at the foundation was the two major trends of "democracy and trade unionism," which emerged as a result of World War I (Miyamoto 1936, 2). Miyamoto was initially willing to break down the social classes and academic cliques (especially law graduates) and develop them into an economic democracy (Sano et al. 1936, 25–29).

In this period, improving the efficiency of management by promoting the rationalization of industry became a social issue to meet the demand for the improvement of working conditions and take measures against the recession in the post–World War I stage. It became an important issue in the industrial competition for national defense after the Washington Naval Conference in 1921. The concept of scientific management, as developed by Frederick W. Taylor in 1911 in his book *Principles of Scientific Management*, had been introduced in Japan.[21] However, it gained attention after World War I. Taylor summarized the key elements of his scientific management theory as "harmony, not discord" and "cooperation, not individualism" (Taylor 1911, 140), attributing it to a "true democracy" bringing about a "complete mental revolution," thus eliminating loafing or "soldiering" voluntarily and enhancing efficiency (Taylor 1947, 29, 217). It can be said that his thinking conformed to Japanese virtues, apart from the core value of political democracy. In Japan, scientific management was developed in the context of labor-management cooperation, based on *onjō shugi* and management-familism, where individual tasks were imposed on the group at the workplace. Even when *onjō shugi* came to an impasse, it did not develop democratically (Hazama 1978, 154–189).

At the end of the Taisho era, the Universal Manhood Suffrage Act was enacted in 1925, granting all men over the age of 25 the right to vote. Meanwhile, the Peace Preservation Law was enacted as a set of laws to punish organizations aimed at reforming Japan's national polity or denying private property. The Japanese Communist Party began campaigning legally under the new election system, but they were arrested under the law across the country in the March 15 and April 16 incidents in 1928 and 1929, respectively. Japan also fell into a severe economic recession for lifting the embargo on the export of gold in 1929 and the influence of the Great Depression that began in New York on Black Thursday in October that year. Among growing criticism from other countries of Japan's cooperative diplomacy toward China, the Kanto Army, a Japanese army unit stationed in Manchuria, caused the Mukden Incident in 1931 to subdue Manchuria, leading to an increase in the domestic political power of the Japanese military. Prime Minister Hamaguchi Osachi was shot and seriously injured over the violation of the emperor's supreme command by the London Naval Treaty in 1930. Moreover, Prime Minister Inukai Tsuyoshi was shot dead by a group of young naval officers in the May 15 incident in 1932. In 1933, Japan declared its withdrawal from the League of Nations over the approval of the State of Manchuria. Over 1,400 young army officers conducted a coup d'état on February 26, 1936, attacking the prime minister's official residence, police headquarters, and other government offices, thus killing ministers and occupying the center of the national government for four days. This coup was taken down by the emperor, but the incident strengthened the army's political power. Thus, the Second Sino-Japanese War broke out in 1937.

As the times changed, some members of the Japanese Technicians' Union approached the Communist Party, and Miyamoto began to harbor fundamental doubts on the idea of trade unionism. He envisioned the union's activities as directed toward national interest, not the social equality of engineers. Accordingly, he came to think it was impossible to adopt the "leading spirit," which competes with nationalism in the international environment of the 1930s. Thus, the statement "the association shall be a professional union of engineers" in Article 1 of the union was deleted in 1930. Moreover, the union changed its name from the Nihon Kōjin Club (the Japanese Technicians' Union) to Nihon Gijutsu Kyōkai (the Japan Technology Association) in 1935, because "kōjin" (technician) meant "coolie" (day laborer) in Manchuria. By changing the board system to a presidential one, the new association changed the policy for promoting technocracy and improving the efficiency of the national economy under the slogan, "guide national opinions based on technology, that is, *gijutsu hōkoku* (repayment for the nation with technology)" (Miyamoto 1936, 2; 1941,

2–3).[22] Based on the Malthusian theory of population, Miyamoto (1937), commenting on a social evolution theory similar to Yamakawa's, stated that "the struggle of mankind from the fundamental problem of depriving each other of goods for survival is fatal." Furthermore, he supported Okochi's science-based policy against capitalism and Prime Minister Konoe Fumimaro's policy emphasizing substantiality and expertise in his first Cabinet from 1937. Okochi also developed a nationalist industry from his science-based policy. Consequently, nationalist policy for the survival of the fittest was strengthened nationwide as the Taisho Democracy weakened.

THE CODE OF ETHICS MODELED AFTER THE AMERICAN PROFESSIONAL SOCIETY

When the Second Sino-Japanese War began in 1937, the Konoe Cabinet began the National Spiritual Mobilization Movement for all Japanese using the slogans: *kyokoku icchi* (national unity), *jinchū hōkoku* (loyalty and repayment for the nation), and *kennin jikyū* (persistent fortitude). The Industrial Patriotic Federation (Sangyō Hōkoku Renmei), organized by the Cooperation Society in 1938, was reorganized into the totalitarian Great Japan Industrial Patriotic Association (Dai-Nippon Sangyō Hōkokukai) in 1940 under the leadership of the Ministry of Home Affairs and the Ministry of Welfare, forcing all labor unions to dissolve. After enacting the National Mobilization Law in 1938, the government assumed complete control of the lives of the populace, thus enforcing industrial mobilization, price control, and rationing. Totalitarian nationalism gained strength across Japan. Nakamura Shunichi of the Tokyo Postal and Telecommunications Bureau, for example, criticized the personnel management of *onjō shugi* as simplistic moralism by referring to Taylor's scientific management. Moreover, he upheld welfare promotion and social justice, modeled on totalitarianism, such as Nazi Germany's National Socialism and the Italian National Fascist Party's Labor Charter of 1927 (Nakamura 1939, 1–32).

Under the circumstance, the JCES (Doboku Gakkai, which corresponds to the Academic Society of Civil Engineering in Japanese) formulated a code of ethics to follow the ASCE case. The JCES, established in 1914, was oriented as an *academic* society of engineering from its inception. The society's charter emphasized that Western scientists and those in other fields in Japan competed to establish academic societies, publish journals, and discuss the results of research and experiments to advance the respective academic fields. Given that the name of the British society was the Institute of Civil Engineers and that of the American society was the American Society of Civil Engineers, the character of the Japanese society was represented through the name

"Engineering," not "Engineer."[23] However, criticism of the JCES' academic-oriented character increased because the engineers' movements developed during the Taisho period, and the international environment changed drastically throughout the 1930s. Miyamoto published "Doboku Gakkai kaizōron" (Reconstruction of the JCES) in 1932 and asked the JCES, which had focused entirely on academic activities, regarding other activities as degraded, to promote "social activities" per the situation at the time (Miyamoto 1932, 62). In the proposal, Miyamoto referred to the ASCE's establishment of the Committee on Development in 1918, and the JCES organized the Committee on Development in 1933. This new policy to shift from a mere academic society to a professional society of civil engineers can be understood as a call to achieve the aim of the Japanese Technicians' Union at the JCES, modeled after the ASCE (Oyodo 1989, 195–205).

The adoption of the "mutual membership agreement (engineering ethics)" was proposed as part of the shift to a professional society (JCES 1933a, 10).[24] Although the JCES was socially recognized, it was not a professional society of civil engineers. While aiming to establish its status under the development policy, it became a problem that "it is regrettable that there is still no mutual agreement among engineers on subjects such as *engineering ethics* in our country" (JCES 1936a, 3), as opposed to the ASCE and the other civil engineering societies (JSCE 1994, 19–20).

The survey for the code was activated in 1936 by the Investigation Committee for the Civil Engineers' Mutual Agreement, chaired by Aoyama Akira, which was organized under the Second Committee on Development. Hirayama Fukujiro, who led the early JCEA after World War II, was also a member. They first collected American and British codes of ethics and practice as a reference. They then limited the subject to the professional conduct of civil engineers and aimed the code at raising their dignity, maintaining their pride and authority, and serving as a guideline for young engineers.

Although each article was based on the secretary's draft regarding the American code, a *doboku kōtoku* (public virtues of civil engineering) was emphasized per the national situation at the time. The committee noted three characteristics of the American code: loyalty to the home country, dedication to the public interest, and respect for their honor (JCES 1938a, 3). These likely came from the ASCE's code, where community welfare, national and local prosperity and development, and public duty were emphasized in the section on their relations with the public (ASCE 1927, 6). While considering the difference between the ASCE values and Japan's nationalistic loyalty and filial piety, the interpretations of public interest and honor were different. In the discussion of the JCES, they confirmed that "the exertion of the Japan spirit is necessary according to the current situation, the project should be national, and the technology should be international." Furthermore, they demanded fairness, cleanliness, and selfless contribution to the country (JCES

1936b, 2). In his 1938 presidential speech, referring to the Western history of civil engineering since Egypt and Rome, Aoyama maintained that it was dutiful to contribute to the development of social and national civilization through civil engineering as a cultural technology (Aoyama 1936). Even Aoyama, who engaged in the construction of the Panama Canal as a foreign engineer, requested to contribute to the nation. Most of the JCES, engaged in Japan's national projects, such as Manchurian development, expected more dedicated services to the nation.

The committee completed the Creed and the Code of Practice of Civil Engineers in 1938. The last report to the president in December 1937 was divided into four categories from a total of fourteen articles: eight on relations among engineers, two on relations between the engineer and the owner, two on relations concerning contractors, and two on relations concerning consultants, which were modeled on the categories of the ASCE code.[25] However, in the final version, published in 1938, they were compiled as eleven articles without any categorization.[26] The final version of the JCES code was less specific than the ASCE code as a mutual agreement for members of a professional association. This result likely stemmed from the organization's academic roots.

In the JCES journal's announcement in 1938, the primary purpose of creating the code of ethics was stated as follows:

> Considering the present situation of the world, it has been in the midst of turbulence and confusion throughout the political, literary, economic, and industrial world since the conventional balance was broken by the Great European War. However, it cannot be denied that emerging nations are obviously developing behind the scene. Developing the future of the nations and ethnicities is an important mission imposed on each national and ethnic group. Especially in the current emergency situation, our country must overcome the present difficulties with all the abilities of the people. So, as civil engineers with a noble mission to contribute to the creation of human culture and take the initiative to found all construction and economic projects, we must be aware that it is urgent to contribute to national society by articulating our position, renewing our insight, and working together to improve this world. (JCES 1938c, 9)

Reflecting this nationalism, Article 1 of the Creed demanded: "The civil engineer shall contribute to the future progress of the state as well as the promotion of the welfare of humankind." Article 1 of the Code of Practice stated: "The civil engineer shall actively serve society for national and public issues with their expertise and experience." Although they referred to humankind and the public without limiting their service to the nation, they were required to serve their country above all else. As the code of ethics was published, JCES volunteers published a claim for the nationalistic contribution as *doboku hōkoku* (repayment for the nation with civil engineering). They appealed as follows:

> Our country is now updating all policies, dealing with the current challenge of top-bottom cooperation, and making every effort to prepare for our empire's future breakthrough. We civil engineers must also recognize the essence of civil engineering and be willing to show our sincerity with our best dedication. (JCES 1938b, 1)

As the war intensified, the mobilization of science and technology proceeded. In December 1938, the Board of Asia Development was established to control the occupation policy toward China. Miyamoto was appointed as the director of the Engineering Department of the Board and many other engineers who had participated in the engineers' movement, such as the Japanese Technicians' Union and the JCES Committee on Development, were also appointed as members of the Engineering Committee on the Board of Asia Development (Oyodo 1989, 258–266). Thus, under this special condition of the wartime regime, engineers obtained the central posts of the national administration. Meanwhile, the Creed and the Code of Practice of Civil Engineers, modeled after the American system, were forgotten along with the increasing demand for contributions and services to the state after the outbreak of the Pacific War between Japan and the United States in 1941.

NATIONALISTIC COOPERATION IN ENGINEERING

In prewar Japan, national survival and development were required to protect the *Kokutai* national polity of the emperor's sovereignty and confront domestic and foreign threats due to modernization and the expansion of international struggles. Thus, not only communism for the revolution but also democracy for popular sovereignty were excluded. Liberalism was inevitably denied, and conservatism became dominant because it became the foundation of the national polity that the line of the Imperial Family has been unbroken for ages. The *Kokutai* pedagogical policy strengthened general moral training to emphasize individual contribution to the country above all. Therefore, ethics education, focusing on the engineering profession, did not develop. In engineering education at the university level, the cultivation of nationalism was required in the 1918 University Order, but the ethics class was not introduced because the object was academic education and research rather than vocational education. The same held for the JCES code, which was modeled on the ASCE. The American-style code of professional ethics did not become popular. The Japanese Technicians' Union, which aimed to improve the social status of engineers via the Taisho Democracy trend, gave up on democracy and reorganized itself to promote technocracy.

From the late 1930s, the institutionalization of science and technology mobilization was promoted to strengthen production capabilities. The shortfall of engineers became a social problem, and the faculties of engineering and technical schools expanded in number. In 1939, Nagoya Imperial University (comprising the Faculty of Medicine and the Faculty of Science and Engineering) and the Faculty of Science at Kyushu Imperial University were newly established, and the faculties of engineering at imperial universities also expanded in size. The industry also supported these expansions. Taga Higher Technical School (currently the College of Engineering at Ibaraki University) was established with an investment from Hitachi, Ltd., and Fujiwara Institute of Technology (currently the Faculty of Science and Technology at Keio University) was established with an investment from the president of Oji Paper Co., Ltd., Fujiwara Ginjiro. Furthermore, to improve engineering education, a proposal was issued to provide practical work activities for students in close cooperation with factories at vocational schools and universities, modeled on the American cooperative course and German cases (Production Control Committee 1938, 8–10, 25–30). In 1940, the Cabinet approved the Science Mobilization Implementation Plan (Kagaku Dōin Jisshi Keikaku Kōryō) to promote scientific research and its practical application for industrial promotion to strengthen the national defense and self-sufficiency by developing resources in Japan, Manchuria, and China.

At the end of May 1941, the Second Konoe Cabinet approved the Guideline for Establishing the New Science and Technology Order (Kagaku Gijutsu Shintaisei Kakuritsu Yōkō), stating:

> To establish a national total war order for science and technology, which is the foundation to complete an advanced national defense state, to promote scientific and technological breakthrough, to stimulate people's scientific spirit as the basis, and to expect to complete Japanese character of science and technology based on the resources of the Greater East Asia Co-prosperity Sphere [Daitōa Kyōeiken].[27]

As a measure of this new order, a proposal was issued to enhance research efficiency by linking basic and applied research per Japan's national situation to promote industrialization research, utilize the research outcomes, and enforce systematic training and deployment of researchers. Against the stimulation of the "scientific spirit" in the people, the low scientific literacy was often considered a national problem. The lack of scientific thinking and its cultivation were considered a deficiency in prewar Japan and became a social issue even after the war.

The attack on Pearl Harbor in December 1941 began the war between Japan and the United States, thereby accelerating science and technology mobilization in both countries. In Japan, the Technical Institution (Gijutsuin) was established in 1942 to promote the previous year's guideline. Inoue Tadashiro and Wada Koroku were appointed as the first president and vice president, respectively. After the war, Inoue became the second president of the JCEA, and Wada became the president of the Tokyo Institute of Technology for the 1944–1952 period and the first president of the Japan University Accreditation Association (JUAA) in 1947. He also arranged for the 1951 Engineering Education Mission to be invited to Japan from the American Society for Engineering Education (ASEE). Accordingly, efforts to improve engineering and engineers' social status were passed down to post–World War II Japan.

The Technical Institution urged the All Japan Science and Technology Association (Zen-Nippon Kagaku Gijutsu Tōdōkai), established in 1940, to organize *neighborhood research groups* (kenkyū tonarigumi) in each related field and systematically mobilized scientists and engineers to undertake joint research for national purposes, thus strengthening science and technology mobilization. The association was integrated into Dai-Nippon Gijutsukai (the Great Japan Technology Society) with Kōseikai and Nihon Gijutsu Kyōkai (previously the Japanese Technicians' Union), which was later reorganized to form the Japanese Union of Scientists and Engineers (JUSE) in May 1946 and took the initiative in the quality control (QC) activities in postwar Japan. The Japan Efficiency Association (JEA, Nihon Nōritsu Kyōkai, currently the Japan Management Association), a leader in production management after the war, was established in 1942. It set forth the maximization of productivity and economic power, efficiency enhancement, and industrial rationalization to ultimately win the war and attain realization by pursuing "efficiency based on Japanese ethnicity" (Morikawa 1942, 1–2; JEA 1942, 65). Regarding the industrial rationalization under state control, pursuits of free economy and corporate profits were denied. Moreover, cost reduction through cooperation between companies and the introduction of scientific management were promoted (Hazama 1963, 175–181). Thus, the science and technology policy was institutionalized in the 1940s in Japan and continued even after the war. Cooperation among the industry, the government, academia, and the military was institutionalized in the early 1940s. Furthermore, strong antipathy to such mobilization became a driver of the postwar anti-industry-academia-cooperation movement in Japan. Chapter 4 discusses the details of the above developments.

In 1943, the Japanese war situation deteriorated further, and the labor shortage became extremely serious. The humanities courses at universities and vocational schools were compressed and converted to science and engineering courses with insufficient preparation, as student mobilization gained strength. Eventually, Japan lost the war on September 2, 1945.

After the war, the movements to improve engineering education and the social status of engineers were again promoted under the supervision of the United States, a democratic country. Although Japan and the United States required engineering ethics before the war, their goals, given their ideologies, differed substantially. Thus, after World War II, Japanese engineers found American engineering ethics as a completely new Western culture. Regarding engineering ethics, chapter 2 describes the postwar engineering education reform in detail, and chapter 3 discusses the postwar improvement of the social status of engineers.

NOTES

1. In Japan, the Education System Order, proclaimed in 1872, introduced the Japanese basic educational system, comprising elementary and middle schools and universities. The elementary school system was reformed by the Education Order in 1879 and Elementary School Order in 1886. Furthermore, ordinary elementary schools offering four-year compulsory education (which extended to six years in 1907) and higher elementary schools were established. Moral training was given top priority in the curriculum development via the Guidelines for the Course of Study for Elementary Schools in 1881. Meanwhile, one five-year ordinary middle school was established in each prefecture, and three-year higher middle schools were established as university-preparatory schools (renamed as the higher school in 1894). Vocational education, including engineering education, also developed as a part of the middle-level education and was institutionalized by the Vocational School Order in 1899. In the educational reforms after World War II, prewar ordinary middle schools were converted to high schools, and prewar higher schools were integrated into the general education program of the postwar university system.

2. The Meiji Restoration is a series of events in which Japan's political structure was converted from the feudal system of the Tokugawa Shogunate to the modern monarchy of the Meiji government. In the nineteenth century, contradictions inherent in traditional centralized feudalism emerged, and Western countries demanded that Japan open to trade and diplomatic relations, particularly from the arrival of Matthew C. Perry, commodore of the U.S. Navy, in 1853. Japan's response was divided into two positions: supporting the shogunate and opening the country and reverencing the emperor and excluding foreigners. Thus, the Tokugawa Shogunate restored sovereign power to the emperor in 1867, and the Meiji era began in 1868. As the Western military powers, with their advanced scientific, technological, and industrial capabilities, forced Japan to open, their introductions into Japan became a top priority for modernization.

3. See chapter 2, section 2.4.

4. Popular sovereignty was denied by the Constitution of the Empire of Japan, established in 1889. The Japanese government was controlled by politicians and bureaucrats on the premise of harmony with its citizens (Banno 2005, chap. 5).

5. Inoue Kowashi, who became the MOE minister in 1893, established the Vocational Education National Subsidy Act and the Engineering Teacher Training Regulation of the MOE in 1894. He also enacted the Higher School Order in the same year and attempted to reform higher middle schools (which had been university-preparatory schools) into professional schools by setting up specialized faculties, such as engineering.

6. Japanese textbooks were initially under government certification but were changed to government designation in 1903. This system had been previously considered, but the policy was changed after a large bribery scandal of adopting textbooks in 1902, where nearly 200 people were arrested and 116 people were convicted.

7. Yamakawa also promoted militarism with moral education based on gentlemanship and Bushido. See Natsume (2017b) for a historical analysis of his militarism.

8. Yamakawa, however, was unable to major in physics in the United States. He enrolled in the civil engineering course at the Sheffield Scientific School for a Bachelor of Philosophy degree (Chittenden 1928, 124–126, 331–338). The school was designed to train practical high-level engineers such as the École Centrale des Arts et Manufactures in France. The undergraduate courses consisted of engineering and other practical fields such as chemistry, metallurgy, civil and mechanical engineering, agriculture, and natural history.

9. Yamakawa ([1905] 1937a, 192) stated that "persisting in the faith of death is the samurai's ambition, but there are many cases where it is not necessary for gentlemanship." Bushido, different from gentlemanship, often demanded that people die to take a path of righteousness instead of living.

10. A popular riot grew out of a rally in Tokyo Hibiya Park on September 5, 1905, against the peace treaty of the Russo-Japanese War. By the time the martial law was lifted at the end of November, 17 people died, 2,000 were injured, and 2,000 were arrested.

11. Spencer objected to nationalist education, because he believed that individual freedom would be respected, and the function of the nation would become restricted in the evolution from a militant to an industrial type of society.

12. Morito Tatsuo, professor at the Faculty of Economics, Tokyo Imperial University, published a paper on the social philosophy of Russian anarchist Pyotr A. Kropotkin in the first issue of the journal of the faculty and was accused of rebellion against the government. Academic freedom, thus, became an issue.

13. The Rice Riots in 1818 caused the expansion of agrarian disputes in rural areas more than labor disputes. Thus, the Nihon Nōmin Kumiai, a nationwide farmers' union in Japan, was organized in 1922.

14. Regarding technical workers, factory laborers became a major social problem during this period. For example, Uno Riemon organized the Society of Industrial Education (Kōgyō Kyōikukai) in 1909 and published many articles. However, this book does not discuss these factory labor issues in detail.

15. After working as the director of the harbor city plans in Tokyo and Osaka, Naoki became the chief engineer of the Imperial Capital Reconstruction Institute

(Teito Fukkōin) and the director-general of the Reconstruction Bureau of the Ministry of Home Affairs, in response to the Great Kanto Earthquake in 1923. He then took a top position among technocrats in Japan.

16. The Civil Service Appointment Ordinance was amended in 1913 as party politics advanced. Moreover, the scope of discretionary appointment was expanded such that party members could become bureaucrats.

17. In his lecture in 1913, Okochi (1919, 1–24) stated that the machine manufacturing industry necessary for the domestic munitions production system had not yet been realized. Thus, he appealed for improvements in engineering education, including practical activities in factories and other subjects necessary for factory management.

18. Yamakawa Kenjiro was appointed as an advisor to RIKEN since its establishment.

19. Okochi proposed a social improvement technocracy to control capital by the labor union of white-collar and blue-collar employees. Oyodo (2009, 32–33) described it as a pioneer in the *technostructure* argument developed by John K. Galbraith in the genealogy of Veblen.

20. For example, the union began a qualification for technicians in 1925 to certify that their abilities were equivalent to those of graduates of higher technical schools for all technicians to access equal opportunities.

21. The Japanese translation was published in 1913.

22. Kōseikai had already changed its aim from improving engineers' social status to the realization of national policy since the National Engineers' Congress in 1924 at the end of the Taisho era.

23. The JCES changed its English name to the Japan Society of Civil Engineers (JSCE) in 1951 (Furuki and Sakamoto 2004, 72).

24. Initially, they shared the Japanese translation of the 1922 code of ethics by the joint committee of the ASCE, the American Institute of Mining and Metallurgical Engineers, the ASME, the AIEE, and the American Society of Heating and Ventilating Engineers, where the "welfare of the Community" was translated nationalistically as "kokuri minpuku" (national interest and people's welfare) at the start of the preamble (JCES 1933b, 12–13).

25. The ASCE code of practice comprised nine articles on relations among engineers, twelve on relations between the engineer and the owner, sixteen on relations concerning contractors, two on relations concerning sub-contractors and material men, and eight on relations with the public (ASCE 1927). The JCES code was simple relative to the ASCE code (JCES 1938a, 2–3).

26. In the reported version of 1937, *gijutsusha* and *gijutsuka* (engineer or technician) were used together, and the title was written as *gijutsuka*, but in the final version of 1938, they were unified to *gijutsusha* (JCES 1938c, 9–10). Although the difference in the meaning of *gijutsusha* and *gijutsuka* is not clear, in Japan today, *gijutsusha* is recognized as a translation of the English word "engineer" and *gijutsuka* is close to technician, sounding like "amateur engineer."

27. *Kagaku Gijutsu Shintaisei Kakuritu Yōkō* [The Guideline for Establishing the New Science and Technology Order], approved by the Cabinet on May 27, 1941.

Chapter 2

Engineering Education and Ethics in Postwar Democratization

INTRODUCTION OF DEMOCRACY UNDER THE ALLIED OCCUPATION

Before World War II, Japanese organizations for engineers emphasized ethics. Ethics was taught as part of popular and vocational education. However, its basic social perspective was very different from that of American professional ethics. Japanese ethics required loyalty and service to protect the *Kokutai* national polity and repay the nation with technology. The JCES adopted a professional and practical code of ethics modeled after American engineering ethics. However, it was foreign and was forgotten as the wartime regime progressed with nationalistic values. Furthermore, universities did not offer ethics courses, because they were specialized in academic higher education. Thus, the systematic introduction of engineering ethics did not progress in prewar Japan.

After Japan's surrender on September 2, 1945, the Allied Occupation drastically changed the nation's social perspective. The General Headquarters, Supreme Commander for the Allied Powers (GHQ/SCAP) enforced the complete elimination of militaristic and ultra-nationalistic features and the democratization of all social systems. As an occupation reform, the Civil Information and Education Section (CI&E) of GHQ/SCAP was in charge of democratizing the entire education system. The Scientific and Technical Division of the Economic and Scientific Section (ESS/ST) of GHQ/SCAP, especially with Harry C. Kelly, was in charge of evaluating and reorganizing the research and higher education system for science and technology, such as the Science Council of Japan (SCJ), a representative organization of Japanese scholars and scientists in all fields of sciences. Kelly's successor, Bowen C. Dees, planned the Engineering Education Mission to Japan by the American

Society for Engineering Education (ASEE) to improve the Japanese engineering education system. Based on the knowledge gained from the international exchange meetings, the Japanese Society for Engineering Education (JSEE, Nihon Kōgyō Kyōiku Kyōkai) was established in 1952.

The JSEE is generally known as the origin of post–World War II Japanese engineering education, comprising all engineering fields. However, before the exchange meetings, Koga Issac, an electrical engineering professor at the University of Tokyo and secretary of the steering committee of the meetings, and other electrical engineering professors had already established the Association for the Advancement of Electrical Engineering Education (AAEEE or A^2E^3, Daigaku Denki Kyōkan Kyōgikai) in 1950 under the guidance of Frank A. Polkinghorn, director of the Research and Development Division of the Civil Communications Section (CCS/R&D). Other engineering fields did not see such early efforts to improve education in telecommunications engineering. The R&D Division and the Industry Division of the CCS introduced American ideas on the industrial development of research and organizational management to the Japanese telecommunications industry. Their instructions on Japanese engineering education reform were part of these activities. Although the CCS did not compel Japanese professors to introduce American engineering ethics, Japanese leaders noticed such ethics when studying American engineering. This chapter describes the process.

INTRODUCTION OF AMERICAN R&D AND MANAGEMENT

Before discussing the issues of engineering ethics and education reform, this section examines the philosophy behind the CCS' introduction of American R&D management, which was a common ground of their policies. The CCS/R&D would have recommended the improvement of engineering education based on the concept.

Air raids by the United States severely damaged many Japanese cities. Much of the telecommunications infrastructure was destroyed; telephone subscribers, which had been 1,100,000 across the country prewar, was now about 540,000 postwar, and production equipment and materials for the reconstruction were extremely scarce (Postwar Communications Industry Editorial Committee 1959, 12–13). Thus, to address these communication problems, the CCS was set up with the establishment of GHQ/SCAP and conducted the reconstruction and restructuring of a nonmilitary and democratic telecommunications industry. Initially, Japan's industrial recovery was restricted to demilitarization and punitive policies throughout the occupation. However, in 1947, this policy was overturned to encourage Japan's economic independence to reduce the financial burden on the United States and make Japan a bulwark against communism per anticommunist policies such as the Truman Doctrine and the Marshall Plan. When the reconstruction of the telecommunications equipment manufacturing industry was completed in mid-1948, the R&D and the Industry Divisions of the CCS provided active support for restructuring the telecommunications industry's development and management system into a more appropriate system from their perspective.

In 1947, the CCS/R&D's director, Raizemond D. Parker, and his staff member, Edmund Gonzalez-Correa, set up a U.S.-Japan joint committee on R&D and recommended to the Ministry of Posts and Telecommunications (MOPT) minister that the Electrotechnical Laboratory (ETL, Denki Shikenjo), a national research institute of the ministry, should be completely restructured. Before the war, the ETL had conducted basic research on power transmission and telecommunications, as well as tests of electrical units and the setting of standards. During the war, mobilization expanded its research scope but neglected economic efficiency (Yoshida 1948b, 3). Given the restructuring, the Electrical Communication Laboratory (ECL, Denki Tsūshin Kenkyūjo) was established in August 1948 by dividing the ETL into telecommunications and electric power branches and integrating the former with four electrical laboratories of the MOE, the MOPT, and the International Electric Communications Co. (Kokusai Denki Tsūshin Kaisha).[1]

The ECL called for democratic services for all people based on economic rationality rather than the ETL's prewar nationalistic and academic-oriented activities. The slogan of the ECL's revolution[2] was *jitsuyōka* (development). Yoshida Goro, the first director, explained it as follows:

> In short, it is how to say *research and development* in Japanese. Although it is said to be *kenkyū kaihatsu* today, *kaihatsu* does not include all *development*. I think that the Japanese word *kaihatsu* captures only a part of it. *Development* has a meaning of *kaihatsu* but does not connote the idea of bringing it up to something concrete and practical. So, it was a difficult task for me to select this word as a slogan for the establishment of the laboratory. (Yoshida 1973, 108)

The Japanese translation of the English word *development* is generally *kaihatsu* or *hatten*, and *research and development* (R&D) is still generally translated as *kenkyū kaihatsu*. However, Yoshida and his colleagues realized through consultation with the CCS staff that *development* was a broader concept than *kaihatsu,* because the former indicated a series of processes that developed research from purpose to a commercially viable industrial product. Therefore, they decided that no word or concept corresponded to *development* in Japan and proposed a new Japanese verbal wording *jitsuyōka-suru* (develop) as a translation (Yoshida 1948b, 3). Namba Shogo, a member of the joint committee with

Yoshida, remembered that they had used the verb *jitsuyō-suru* (utilize) but had not used the verb *jitsuyōka-suru* (develop) before (Namba 1973, 398).

However, as early as 1935, Okochi Masatoshi of RIKEN used the verb *jitsuyōka-suru* along with *jissaika-suru* (realize) and *kōgyōka-suru* (industrialize) to refer to inventions and research.[3] Moreover, the Japan Society for the Promotion of Scientific Research (JSPSR, currently JSPS) commented that "without doing *jitsuyōka* and *kaihatsu*, they [basic and applied research] will become a waste of precious resources" (JSPSR 1939, 126).

The Foundation for the Promotion of Scientific and Industrial Research of Japan (FPSIRJ) was established in 1932 when Japan's research grants continued to be reduced due to the economic recessions of the early Showa era (1926–89). Sakurai Joji, a chemist and the president of the Imperial Academy, and others stated in the foundation's charter that they conducted the academic promotion as an "executive agency for cooperation" between the nation and the people because of a lack of funds to achieve research results, lack of facilities and collaborations to industrialize them, large companies' dependence on import and imitation, and people's low level of understanding and interest. Meanwhile, Germany, Britain, and the United States were developing national power by promoting the industrialization of research (JSPS 1998, 321–322). The draft of this charter referred to the specific models of *jitsuyōka*, the Carnegie Foundation, the Rockefeller Foundation, and Bell Telephone Laboratories in the United States together with the Kaiser Wilhelm Society in Germany (FPSIRJ 1932, 7). Nevertheless, in the run-up to its establishment, the original Japanese name had been Gakujutsu Sangyō Shinkōin (the Institute for the Promotion of Science and Industry). However, the MOE changed the name to Nihon Gakujutsu Shinkōkai (the Japanese Society for the Promotion of Science), omitting the "Industry" when requesting the budget. Although the original English name had been "the Foundation for the Promotion of Scientific and Industrial Research of Japan," it was also changed to "the Japan Society for the Promotion of Scientific Research" in 1936 (JSPS 1998, 16).[4]

According to Nagaoka Hantaro (1938, 9), chairperson of the academic section, it became the main purpose of the JSPSR to provide various research grants for natural sciences and comprehensive research (sōgō kenkyū) for national purposes, including social sciences and humanities. However, the *jitsuyōka* to the industry remained insufficient because they could not set up development laboratories like in Europe and America. The JSPSR had classified research such as "purely academic; basic applied; and development, industrialization, or practical research" (JSPSR 1941, 539). While the extent to which economic rationality was sought was vague, it was similar to the postwar understanding. However, the idea was not widely accepted, and the system was hardly realized.

Even the prewar ETL organized the Society of Utilization of Electrotechnics (Denki Gijutsu Jitsuyōka Kyōkai) in 1942 to manage patented technology, industrializing research, and awarding grants (dissolved in 1948). However, the *jitsuyōka* was generally limited to practical application or utilization. The term *jitsuyōka* was not widely used; they could not systematically combine research activities with development.

The ECL's R&D was modeled on the Bell Telephone Laboratories (Yoshida 1965). Most of the CCS staff consisted of people involved with the Bell System (Adams and Miranti 2011, 133–135). Both Parker, director of the R&D Division, and his successor, Frank A. Polkinghorn, who arrived in Japan in August 1948 when the ECL was established, came from Bell Telephone Laboratories. The organization cooperated with the armed forces in various ways during the war and continued cooperating with GHQ/SCAP in civilian communications R&D after the war. Polkinghorn's dispatch to Japan was offered by President Oliver E. Buckley and Executive Vice President Mervin J. Kelly when Parker's wife became ill. It was canceled once due to concerns for his family (with four young children) and Parker's decision to extend his stay. However, six months later, Parker obtained another job in the United States due to his wife's illness, and Polkinghorn agreed to be sent to Japan to replace him (Polkinghorn 1969, 3).

Polkinghorn defined *development* as "to produce a practical embodiment of an idea" (Polkinghorn 1949c, 1) and noted in his 1949 speech that there was no appropriate Japanese word to express its meaning:

> Japan has a few outstanding scientists that rank well with scientists in other parts of the world. The Japanese have long been known for their ability to copy products devised elsewhere and when they take the care to do so they are capable of producing things of high quality. A great weakness seems to lie in the lack of the various types of engineers and scientists required to fill the gap between these two extremes. I am told that the Japanese have not, until recently, had a word for development, the process of taking an idea and carrying on all the engineering necessary to put equipment based on this idea into production. This lack of a term implies a lack of understanding or appreciation of the importance of this phase of the work. (Polkinghorn 1949a, 3)

They thought the concept of development was absent in Japan because of Japan's lack of understanding and the considerable change in the understanding of "development" in the United States during the early twentieth century. During this period, the establishment of corporate laboratories, such as Bell Telephone Laboratories in 1925, increased, and technological research shifted from individual invention to organized and collaborative development. With the organized industrialization of product development, the discussion of *development* for research began in the 1920s and was classified especially in *Scientific Research and Social Needs* by evolutionary biologist Julian Huxley in 1934. Thus, the U.S. government established the Office of Scientific Research and Development (OSRD) in 1941. Hence, the abbreviation "R&D" became prevalent, and the National Science Foundation (NSF) and the Organisation for Economic Co-operation and Development (OECD) has promoted its index analysis since the 1950s (Godin 2017, 83–98).[5]

Polkinghorn argued that cooperative group effort was indispensable to modern research because "Research and development can no longer be based on the knowledge and effort of a single individual," like in the days of Leonardo da Vinci and Benjamin Franklin. He also emphasized the need for management: "If research and development have become a cooperative effort, it follows that suitable management will be required to organize and direct the effort of the group to the desired end" (Polkinghorn 1949c, 2). Furthermore, in the formal opening remarks of the ECL, Polkinghorn emphasized the importance of morale and pride for organizations, as well as cooperation and teamwork. Thus, with reference to the ideal or motto of the Bell System, he called for economic and democratic *service* as follows:

> That means the ideal of giving the greatest service to the greatest number of people at the least cost. It is an attitude of mind that fosters thinking of all problems in terms of their ultimate value to humanity. I can recommend nothing better than that you adopt that motto as your own. (Polkinghorn 1949b, 219)

The Bell System introduced Theodore Vail's *universal service* program in 1907 and evolved under the management philosophy of *One System, One Policy, and Universal Service*. Polkinghorn's analysis found that the Japanese telecommunications industry could not realize a better telephone system primarily, because the excessive privilege of government and companies had hindered universal services for many people (Polkinghorn 1950a, 2–3). This emphasis on cooperation for democratic management was similar to Frederick W. Taylor's notion that his scientific management was necessary for "all great things" in his *Principles of Scientific Management* in 1911 (Taylor 1911, 140–141).[6]

The ECL also called for a democratic and dynamic management system, modeled on the relationship between AT&T, Bell Telephone Laboratories, and Western Electric Manufacturing Company, with equal coordination between the divisions of service, research, and development by introducing an economic assessment (Yoshida 1948c, 14). Regarding the close coordination and cooperation between them, a development system of research was required (Yoshida 1948b, 3). Engineers were required to provide services even in prewar Japan. However, they were nationalistic services and cooperations, not democratic services through economic activities. Although democracy could also be oriented toward the nation in general, its fundamental values were different from those of prewar Japan. Taylor's scientific management had been introduced in Japan, but they could not transform each local improvement into an overall harmonious improvement. The prewar system's efficiency was inconsistent, and many Japanese products were not of high quality before the war. The quality deteriorated even further during the war. Based on such reflections, Japanese engineers sought democratic and autonomous management of their industry.

The CCS/R&D was separated from the Industry Division in 1947 and reintegrated in 1949. These two divisions jointly supported the improvement of the management systems of the Japanese telecommunications equipment industry. While the R&D Division guided development to research institutes, the Industry Division began a seminar (the Course in Industrial Management) in 1949 for telecommunications equipment companies to improve their management as the next step in Japan's reconstruction.

Shortly after the war, a shortage of production equipment, as well as a reduction in working hours due to food shortages, reduced the production yield to only 6 percent. Vacuum tube shortages caused frequent breakdowns at telephone relay stations (Editorial Committee for the Unofficial Communications History 1962, 468). Thus, to counter these shortages, the Japanese Union of Scientists and Engineers (JUSE), founded in 1946, launched the Quality Control Research Group in 1948 to promote the introduction of statistical quality control (SQC). SQC, proposed by Walter A. Shewhart in 1924, was introduced in the Bell System; Shewhart conducted the research at Bell Telephone Laboratories. Homer M. Sarasohn (Radio Equipment Supervisor of the CCS Industry Division, an engineer from Raytheon Manufacturing and the MIT Lincoln Radiation Laboratory) estimated that SQC was too theoretical for the current Japanese situation and believed that the poor working environment and management system should be radically improved first.[7] Thus, he organized a seminar, "Management Training Course for the Communications Manufacturing Industry," with his colleague Charles W. Protzman (a telephone equipment engineering advisor of the CCS Industry Division and an engineer from Western Electric) and Polkinghorn at the R&D Division. The CCS management course, which started in September 1949, was very successful and continued until 1974. It was reputed to be an origin of Japanese postwar corporate management.[8]

The management course introduced a wide range of fundamental management ideas, from corporate philosophy on purpose, policy, and social responsibility to the scientific management of productivity, total quality control (TQC), cost control, and internal control, based on the American-style top management and governance with outside directors. Sarasohn and Protzman explained the reason for planning the seminar: "The new democratic society of Japan, where *private* enterprise and *free* competition are the rule, will require that management functions upon the basis of a modern set of concepts" (CCS/R&D 1949). In those democratic concepts, they emphasized *cooperation*. When the course textbook was published, Polkinghorn wrote the following in the preface:

> There are certain occidental basic concepts behind the practice of western industrial management which, while they may be obvious to the occidental mind, may not be clear to the oriental mind. For this reason it would appear advantageous before starting the book and becoming immersed in the detail to state in simple terms some of the fundamental sociological and psychological concepts underlying the material of this book. Democratic western civilization recognizes that each person has equal fundamental rights with all other persons and that the dignity of the individual is the foundation upon which society is built. [. . .] It recognizes that although individuals have life, sorrow, happiness, the things for which humanity strives, nevertheless they have greed, selfishness and other antisocial characteristics which must be controlled either by the individual or by society. For this reason, cooperation is to be regarded as a primary virtue and fostered in every way possible. (Polkinghorn 1952, 7–8)

A later interview with Polkinghorn noted how "he *must* stress the contribution democratic practices made to successful management" (Hopper 1982, 26) in this preface.

The seminar was aimed to introduce American management theories to Japan. However, their ideas differed from those of American management philosophy, reflecting their unique perspective of Japanese culture. For example, Sarasohn emphasized a spiritual and hierarchical human relationship between leaders and followers, such as respect and appreciation, and advised: "A leader's main obligation is to secure the faith and respect of those under him" and the leader "must himself be the finest example of what he would like to see in his followers" (Hopper 1982, 24). Kenneth Hopper, an engineer who had studied postwar American and Japanese manufacturing cultures, concluded that the CCS' activities were motivated by a disinterested desire to create a better Japan that stemmed from old-fashioned Puritan zeal. He explained that the reason was "dedication that went beyond the call of duty and was certainly not reflected in the money, which they earned while there or the rewards they would receive on their return home" (Hopper and Hopper 2007, 111).[9]

In the United States, *development* had become a part of R&D since the 1920s. With the outcomes of scientific research being industrialized, the systematic understanding of R&D had progressed remarkably. Thus, the very meaning of *development* had been altered to emphasize the managerial and economic perspective of organizing scientific knowledge. As it was based on the idea of democracy, the implications differed from those of the nationalistic promotion of science and industrial rationalization in prewar Japan. After the war, when comparing their research outcomes with those of the United States, given their defeat, the CCS' democratic perspective was quite acceptable to Japanese telecommunications engineers. Furthermore, the asymmetry of power, the weakness of occupied Japan, prevailed, and it was easy to agree to any reform for the absolute goal of reconstructing Japan. In this special kind of situation, Japanese engineers tried to improve telecommunications engineering voluntarily, based on new democratic and economic ideas. The next section discusses how CCS staff and Japanese engineering professors improved engineering education based on these ideas.

THE CCS-LED IMPROVEMENT OF ENGINEERING EDUCATION

As a part of the reconstruction of the Japanese telecommunications industry, the CCS worked on engineering education reform, as well as R&D and management. In a preliminary survey conducted by the Occupation Force engineers shortly after the war, Japanese supervisors always spoke about the inadequate training of most employees for positions in telecommunications enterprises (GHQ/SCAP 1990, 78). First, the Japanese proposed to integrate the MOPT's training schools into a telecommunications college. However, GHQ/SCAP instructed the telecommunications enterprises to provide only vocational education and leave general education to the MOE. Per this policy, the Telecommunications Personnel Training Act was enacted in August 1948, and Suzuka Telecommunications Training School (Suzuka Denki Tsūshin Gakuen), an in-house vocational school for telecommunications technicians, was established in June 1949 under the guidance of James M. Roche and John L. Vandegrift at the Telephone and the Telegraph (T&T) Division of the CCS.[10]

In addition to establishing the training schools for telecommunications technicians, the CCS/R&D also encouraged the improvement of engineering education at universities. Polkinghorn, who came to Japan in August 1948, unofficially assembled Sakamoto Toshifusa, Koga Issac, and other engineering professors in Tokyo, through Yoshida Goro's intermediation, and exchanged opinions on the problems of Japanese engineering education in the office of Director Yoshida at the ECL. Yoshida told Sakamoto: "This time, it is your turn" (Koga 1973, 242–243).

Polkinghorn estimated that Japanese engineering education lacked the methodology of R&D management. According to him, Japanese engineers were quite ignorant in their consideration of economics, financial costs, and how to apply their knowledge. Many professors were intelligent, had "a good book knowledge," and practical experience in the MOPT or telecommunications industry, but they were not interested in vocational education and delivered lectures that were too academic and theoretical at a high level without consideration for practical applications. Furthermore, due to the loose lecture system and lack of textbooks, students could pass without much effort, because there was no record of attendance, few questions in class, no homework assignments, and only one or two examinations a year. Therefore, Polkinghorn requested Japanese engineers to train their understanding of economic and management factors as an essential part of engineering education and called for more discipline on the part of engineering professors to improve the educational impact (Polkinghorn 1949a, 1950b). In response to Polkinghorn's claims, Japanese engineering professors expressed skepticism about the American request and noted inconsistencies with Japan's existing university system. However, from the students' perspective, they did not need to maintain their current regulations. Thus, they finally decided to address the highlighted problems (Koga 1952, 36).

The CCS began to address Japanese engineering education issues, delegated by the CI&E at a meeting with Walter C. Eells[11] at the CI&E and Vandegrift at the CSS/T&T. Initially, fourteen lectures by the CCS were held every Tuesday from February to the end of May 1949. Table 2.1 presents the subjects of each lecture.

These titles show that the practical issues of R&D and management, described in the previous section, directly became issues of engineering education. For example, the first lecture discussed how Japan did not distinguish between quality control and material inspection, which did not lead to product improvement (Polkinghorn 1969, chap. 11, 4). Polkinghorn delivered four lectures, including two on "engineering economics," which dealt with capital, instead of the governmental and *zaibatsu* conglomerate treasury, as well as depreciation and cost studies.

Following the lectures, Japanese electrical engineering professors began an organized effort to reform their engineering education in 1949. They held a summer course at Suzuka Telecommunications Training School from August 14 to 20, with twenty-nine participants from eleven universities across Japan. Polkinghorn and Vandegrift also delivered lectures. At the next summer course at Suzuka in 1950, seventy-two professors from forty-four of the forty-nine universities under the postwar system participated; they decided on concrete plans and the name of their association: the AAEEE.[12]

Table 2.1 Lectures on Engineering Education by the CCS (February 8 to May 31, 1949)
Sources: Data from Polkinghorn (n.d.b); Koga (1952, 36–38)

	Date	Subject
1	February 8	Sarasohn: Inspection and Quality Control
2	15	Sarasohn: Continuation of Talk on Inspection and Quality Control
3	March 1	Eells (CI&E): New Methods of Education
4	8	Gonzalez-Correa: Nature & Organization of R&D
5	15	Visit to Tsujido Laboratory of the ECL
6	22	W. L. Wardell (Delivered by Polkinghorn): Sociological Aspects of Electrical Communications
7	29	Polkinghorn: Engineering Economics I
8	April 5	L. Q. Moss (CI&E): Psychology as Applied to Engineering Education
9	19	Polkinghorn: Engineering Economics II, Applications
10	26	Polkinghorn: On Japanese Export Goods
11	May 10	Polkinghorn: Review & Supplement of the Foregoing Lectures
12	17	Eells (CI&E): Teaching Methodology, Marking Methods, and Measuring Teaching Ability (the Purdue system of student rating their professors)
13	24	Report by Professors on the Progress Made in Adopting Ideas in their Particular Universities
14	31	S. Nishiyama (CCS) & M. Abe (ECL): Concluding. College Life in America

Polkinghorn believed that the Japanese social perspective had a fundamental problem for engineering education, which he maintained at the lectures at seven major universities across Japan: "The development of an adequate telecommunications system for Japan depends upon obtaining men with suitable broad social and engineering concepts as well as scientific knowledge" (Polkinghorn 1950b, 1).[13] Polkinghorn criticized the Japanese social perspective: "They must not only have modern scientific knowledge but must also have an adequate sociological point of view." He stated:

> The civilization of Japan, while it may appear to be dynamic, and undoubtedly has made great strides in the past century, retains much of the static philosophy which characterized Japanese life up to the Meiji period. The civilization of Japan is essentially one that protects the position of people of various classes rather than one that fosters the growth of all individuals in accordance with their imagination, ability, and industriousness. There is little in the normal Japanese life of the friendly rivalry and competition that has been common in the field of sports throughout the world and which encourages each individual to do his utmost. (Polkinghorn 1950b, 2)

Relative to a dynamic democratic society, Polkinghorn thought Japan had various outdated and static constraints: "The feudal society which has existed in Japan has prevented the rise of many people of ability and retarded Japan's progress" (Polkinghorn 1949a, 4). Thus, for telecommunications organizations to take advantage of engineering, he concluded that "Japan needs men with modern democratic thought and dynamic action." The university had a duty to develop engineering education based on social and economic competitive perspectives (Polkinghorn 1950b, 4).

Koga agreed with Polkinghorn's argument and transliterated *engineer* as "enjinia" (with Japanese katakana characters) in his translation of Polkinghorn's lecture because the meaning of the English words *engineer* and *engineering* implied social welfare and management, which would be lost when translated into the well-known Japanese words *gijutsusha* and *gijutsu* (Polkinghorn 1950b, 1). In his translation's note, Koga referred to the following definition of *engineer* by Compton, based on information from Sarasohn:[14]

> An engineer is a person who harnesses powers of nature to contribute to social welfare by using materials via the application of mathematics, physics, biology, and economics through the help of observations, experiences, and scientific discoveries and inventions. Engineers are different from technologists or technicians because they have to be interested in the functions of the organization they belong to, as well as the technical aspects of their works. Moreover, they have to be involved in the overall economic and managerial aspects. (Polkinghorn 1950b, 1)

Polkinghorn stated that "Japan has shown herself to be better at fundamental science than she has at engineering, the process of adapting science to the needs of mankind" (Polkinghorn 1950b, 4). Furthermore, he recommended that, for the sake of Japanese students, good teaching, like in the United States, should be imperative and relate thesis work with practical activities rather than academic ones, rendering professors' research activities to be secondary.[15]

Reflecting on the problems of education and research that disregarded the methodology of development, Koga stated the following regarding the prewar ETL as a typical case:

> Some of us were familiar with the situation in Europe and the United States through visits; we had considerably tried to introduce the merits of American college education, but they were insufficient because not only our responsibilities but also the way society led graduates were inadequate. In particular, since companies and government offices tended to emphasize pure scientific research

and treated only those who engaged in it to be excellent, young people competed only in that direction; so, the practical problems (not restricted to technical problems) that engineers should have examined in their workplace tended to be underestimated at any rate. The ETL, before the reorganization, was regrettably the outstanding example. (Koga 1952, 33)

This reflection shows the direction of democratic engineering education that Polkinghorn championed. The suitable social concepts for Polkinghorn were represented by the idea of service to humankind, which was closely related to democracy and his emphasis on *development*.

Polkinghorn's democratic perspective of society also shaped the ethical norms of engineers. He described *noblesse oblige* as "a responsibility to serve mankind in proportion to their ability and position" and emphasized it as "a considerable factor in the motivation of western professional men" (Polkinghorn 1950b, 1). He also argued that company executives and engineers should foster mutual confidence, and capable people should obtain a position to use their maximum abilities in a democratic society (Polkinghorn 1949a, 3–4). He stated: "This idea of service to mankind is a basic social concept behind western progress which is often lost from sight" (Polkinghorn 1950b, 1).

Polkinghorn's understanding of *noblesse oblige* was premised on a democratic society. Even in prewar Japan, *noblesse oblige* was also described as a Bushido value. Nonetheless, it was a feudal concept, excluding service to citizens or humankind. Nitobe Inazo (1905, 4) described Bushido as "the precepts of knighthood, the *noblesse oblige* of the warrior class" and introduced a democratic Heimindo (the Citizens Way) in 1919 to extend Bushido to the general public without class discriminations. However, it was never accepted in Japan. This historical alteration in meaning was similar in the West. In nineteenth-century France, after the civil revolution, the term *noblesse oblige* was used to connote services to the general public without class restriction, as Polkinghorn claimed.

Thus, Polkinghorn emphasized *cooperation* toward the *service*, as echoed in the opening ECL remarks. Moreover, he called for industry-academia cooperation to improve engineering education. Polkinghorn called on academia and industry groups to organize the Telecommunications Industry Education Meeting (Tsūshin Kōgyō Kyōiku Kondankai) with board members of the Federation of Japan Electric Communications Industrial Association (FJECIA)[16] and telecommunications engineering professors at five major universities in the Tokyo area (the University of Tokyo, Tokyo Institute of Technology, Keio University, Waseda University, and Nihon University) in March 1950 to bridge the gap between universities and industry.[17] The main topics included training production engineers, moral education, and training engineers after employment (Polkinghorn, n.d.a).[18]

Quoting the same definition of *engineer* in his translation's note on Polkinghorn's lecture as "a true engineer," Koga and Taki Yasuo, attendees of the meeting, reported that universities should cultivate economic ideas and afford the highest priority to character-building: "The foundation of education hinges on the completion of personality [...] it is fundamental for people as well as engineers, even in university education, that we should do our best for the completion of personality" (Koga and Taki 1952a, 2). As part of the character-building, they were convinced that there were specific professional ethics to which engineers should adhere through their duties, and practical training through specific examples rather than abstract concepts would lead to the "completion of personality," which they believed was the fundamental purpose of education.[19] They stated the following about the cultivation of professional ethics:

> Even though they are fundamentally summarized into one phrase, completion of personality, engineers should have a specific moral to follow. It seems to be necessary to train this through each specific case, not just by abstract concepts. For this purpose, something like the Canons of Ethics for Engineer adopted by the AIEE of the United States is very helpful. Also, in our country, it is desirable to create such a thing that is appropriate for the nation. (Koga and Taki 1952a, 2)

They translated the Statement of Principles of Professional Conduct of the AIEE into Japanese, which was adopted in August 1950 and published in the February 1951 issue of *Electrical Engineer* (Koga and Taki 1952c, 3–6; AIEE 1951). It comprised three sections: the Canons of Ethics for Engineers, the Code of Business Practice, and the Use of Membership Designation and Institute Insignia. The first was the same as the Canons of Ethics for Engineers developed and promulgated by the ECPD in 1947 and the second was inherited from their original code, the Code of Principles of Professional Conduct, adopted by the AIEE in 1912. Among such values, Koga and Taki emphasized *cooperation* as follows:

> Furthermore, for engineers, the spirit of cooperation is especially indispensable, and it is necessary to focus on its cultivation. Cooperation seems to be one of the most important factors for the development of technology in the future since technology is never achieved by the power of a single individual but by the cooperation of many people. (Koga and Taki 1952a, 2)

The emphasis on cooperation in professional ethics was quite similar to Polkinghorn's understanding. It is not clear how democratic Koga assumed the idea of cooperation was, but it is clear that he was actively adopting postwar democratic educational philosophy. He translated Polkinghorn's "social education" as "general education" (ippan kyōiku) per the terminology of the postwar educational reform based on the American system (Polkinghorn 1950b, 1). Affirming the necessity of taking advantage of the new university system, Koga and Taki stated that the system "aims to create better-completed graduates than before, and its focusing on general education is just for the purpose." Furthermore, they added that "its purpose for engineers is to create people with broader knowledge and greater potential for future growth" (Koga and Taki 1952a, 2). Koga understood engineering ethics education to be part of the liberal arts education introduced in the postwar system. Meanwhile, he also called for cooperation between universities and industry, because he saw ethics education as part of professional education rather than just liberal arts or civic education (Koga and Taki 1952a, 1–3).

From this democratic perspective, engineering education was promoted in the telecommunications field, especially under the guidance of Polkinghorn at the CCS/R&D, thus raising awareness of engineering ethics. Polkinghorn was willing to cooperate with Japanese engineers. He also restored the communication between the Institute of Electrical Engineers of Japan (IEEJ) and the AIEE, which had been interrupted by the war, and received a letter of appreciation from the IEEJ in January 1951. Thus, to show gratitude to Polkinghorn, electrical engineering professors presented him with autographs at the Joint Conference of the IEEJ, the Institute of Electrical Communication Engineers of Japan (IECEJ), and the Illuminating Engineering Institute of Japan (IEIJ) in May 1951 (see figure 2.1).

INDUSTRY-ACADEMIA COOPERATION IN ENGINEERING EDUCATION

In postwar engineering education in Japan, *cooperation* was particularly emphasized, and engineering professors established the JSEE in 1952 with an expectation of industry-academia cooperation. Later, in the 1990s, the JSEE promoted engineering education reform and the introduction of engineering ethics by establishing JABEE. In the 1950s and 1960s, the phrase *sangaku-kyōdō* (industry-academia or industry-university cooperation) was used instead of *sangaku-renkei* (industry-academia or industry-university collaboration), which is common in Japan today. The phrase *sangaku-kyōdō* abstractly meant a partnership between university and industry and more specifically implied "co-operation between the various industries of the city and the engineering college of the University" (Schneider 1906, 7), which Herman Schneider, dean of the College of Engineering at the University of Cincinnati, reported as part of the technical education. This method was a "sandwich system" to improve students' understanding of engineering

Figure 2.1 Autographs of Members of the AAEEE at the Joint Conference of the IEEJ, the IECEJ, and the IEIJ in May 1951. "Original Autographs: Mr. F. A. Polkinghorn, 4th, May, 1951." box 18, folder 20, 21, 22, May 4, 1951, Kenneth Hopper Papers on Management, Drucker Institute. http://ccdl.libraries.claremont.edu/cdm/search/collecti on/khp/searchterm/autographs/

through vocational training by systematically alternating periods of on-site working at a company and studying at college every few months. It was also a "learning system with earning" to support students' financial independence (JPC 1958, 39). In the United States, many universities recognized this method's effectiveness and promoted its systematic introduction.

This method was introduced to Japan under the occupation as an excellent method from the United States. As described in chapter 1, practical training was emphasized at the Imperial College of Engineering (the predecessor to the Faculty of Engineering at Tokyo Imperial University) and Osaka Higher Technical School in prewar Japan. The program was also described as a "sandwich" program at the latter, but there was no on-the-job training at private companies in cooperation with the industry. Japanese universities emphasized academic education rather than vocational education, requiring a graduation thesis rather than practical training.

The GHQ-ESS/ST established the Japan Association for Science Liaison in 1946 to reorganize Japan's national scientific organizations. However, it was only for scientists. Thus, the ESS/ST asked Komagata Sakuji (director of the ETL), Kaneshige Kankuro, and Koga to prepare another organization for engineers, and the Japan Association for Engineering Liaison was established four months later to reform Japan's engineering systems (Dees 1997, 104–109, 275–278). These reforms, including the establishment of the SCJ in 1949, were led by Harry C. Kelly of the ESS/ST. After Kelly's return to America in 1950, his successor, Bowen C. Dees, encountered the challenge that professors at Japanese national universities did not have a reliable way to apply their knowledge to the industry because they, as national public officers, were prohibited from receiving remuneration from companies. Industry-academia cooperation means new academic freedom for professors at national universities. Therefore, Dees proposed to Wada Koroku (president of the JUAA) and Kameyama Naoto (president of the SCJ) to invite the ASEE's engineering professors. Furthermore, they decided to hold the exchange meetings of the Institute for Engineering Education (IFEE) in major cities across Japan, supported by the ASEE and the Unitarian Service Committee, Inc. of New York (Dees 1997, 278–283; Sasaki 1951, 27; 1962, 30–31). Wada chaired the IFEE, and Koga was a secretary of the steering committee.

Fifteen members of the Engineering Education Mission, chaired by Harold L. Hazen of MIT, visited Japan from July to August 1951. The site visits and exchange meetings were held in Tokyo, Kyoto, Kobe and Osaka, Fukuoka, Nagoya, Sapporo, and Sendai, with 1,455 people from the industry, academia, and the government (the MOE and the Agency for Industrial Science and Technology, AIST). In the opening speeches from the Japanese, Amano Teiyu, the MOE minister, and Nanbara Shigeru, president of the University of Tokyo, strongly appealed for a close industry-university relationship. The American professors also emphasized the importance of close and flexible relations between teaching and research, students and faculty, and industry and engineering universities. They reported that high school graduates were deprived of opportunities for sub-professional education required by the industry because specialized schools under the old system were transformed into universities by postwar education reform (Japanese Central Executive Committee 1951; Technical Education Division of the MOE 1952).[20] In Japan, the Act on Promotion of Vocational Education was enacted in June 1951 because of the problem created by the postwar promotion of general education, which reduced vocational education opportunities. Although junior high schools through to universities were subject to vocational education under this law, universities without specific provisions were not subject to its enforcement. Regarding university education, the transition to the new system and the introduction of general education caused concern about the decline in the knowledge and skills of graduates for work in engineering (Oyama 1952, 92–93; Tsuzaki 1951, 1).

The JSEE, modeled on the ASEE, was established in 1952, following the strong suggestion of the Engineering Education Mission to Japan. The JSEE placed particular emphasis on industry-academia cooperation, aiming "to contribute to the development of the Japanese industry through intimate collaboration between universities and industries and universities concerning engineering and the promotion of engineering education at the university" (Article 3 of the Regulations). The establishment was prepared by engineering professors, particularly Shimizu Kinji (president of Nagoya Institute of Technology),[21] Sasaki Shigeo (Tokyo Institute of Technology and the JUAA), and Koga, in cooperation with the industry and the MOE.[22]

Furthermore, in June 1956, the Council for Industrial Rationalization (Sangyō Gōrika Shingikai) of the Ministry of International Trade and Industry (MITI) published an advisory report, *On the Industry-Academia Cooperation Education System* and proposed practical methods for industry-academia cooperation, describing it as a factory training system and "a measure to improve the quality of people who supply industry, especially college engineering students supposed to be at the center" (Council for Industrial Rationalization 1956). This advisory report popularized the term "sangaku-kyōdō" (industry-academia cooperation) (Hiroshige 1961, 102).[23]

Moreover, the Japan Productivity Center (JPC), established in 1955, dispatched the Industry-Academia Cooperation Expert Visiting Team (Sangaku-kyōdō Senmon Shisatsudan) to the United States in 1957 as a part of its overseas dispatch project. Eleven JSEE professors from across Japan participated in this tour, including Shimizu (president of the JSEE) as the chairperson and Okoshi Makoto (engineering professor at the University of Tokyo) as the secretary-general (JPC 1958). The industry-academia cooperation in the visiting team's report covered not only education but also research.

The Industry-Academia Cooperation Committee (Sangaku-kyōdō Iinkai), involving four or five members from each industry, academia, and government, chaired by Ishikawa Ichiro, deputy chairperson of the Japan Atomic Energy Commission, was organized to respond to the report. From the academic community, Uchida Shunichi (president of Tokyo Institute of Technology), Ohama Nobumoto (president of Waseda University), Kaya Seiji (president of the University of Tokyo), and Shimizu participated. The committee recommended that industry-academia cooperation was the most effective way to realize an "advanced industrial nation" and proposed the following: (A) research cooperation between universities and industry, (B) a system where industrial engineers constantly studied and applied advanced science (continuing education), and (C) training of excellent engineers in both academic subjects and practical skills. The committee also called for the industry to support both education and research in universities "because universities are an important source of high-level engineers who will contribute to industry" (Industry-Academia Cooperation Committee 1959, 15–17).

Industry-academia cooperation became a general university policy for Japan's economic growth beyond engineering education. The generalization was apparent in the committee's conclusion:

> Reflecting on the current situation of our country, we can find many deficiencies in realizing an advanced industrial nation, and we need to make efforts to realize it through industry-academia cooperation and transform from imitative Japan to creative Japan as soon as possible. (Industry-Academia Cooperation Committee 1959, 17)

Thus, to achieve this goal, the committee first requested MOE in 1958 to create sponsored scholarships as an "industry-academia cooperative system for continuing education after graduating from university" (Industry-Academia Cooperation Committee 1959, 17–20). Furthermore, the contract researcher system (jutaku kenkyūin seido) at national universities was introduced as a reeducation system for engineers in the same year. Expanding the theme from engineering education to national industrial policy, the idea of industry-academia cooperation became an abstract concept.

Although it became a rare case, given the expansion of meaning, there were attempts to introduce the industry-academia cooperative system to engineering education. As a remarkable example, Toyo University established the Faculty of Engineering in 1961 for engineering education modeled on the American-style sandwich program. This establishment was made possible by a donation of 200 million yen from Hitachi, Ltd., with the cooperation of Kurata Chikara, president of the company. The faculty's preparation and the selection of faculty members were led by Okoshi Makoto, who was invited when he was a senior researcher at RIKEN. Okoshi emphasized the significance of introducing the American industry-academia cooperative system since participating in the JPC's visit in 1957. His request was accepted; thus, it was expected to be introduced throughout the faculty (Toyo University 1994, 416–426, 473–483). Moreover, Daido Technical Junior College, established in 1962, introduced a sponsored scholarship following the industry-academia cooperative system (Nishikiori 1963).

However, these cases were exceptional and not popularized. The case of Toyo University was also far from ideal. Some universities could only make short-term off-campus training compulsory, but most of these were not approved for formal credits because of challenges in grading (JSEE 1965, 60–64). The industry avoided the sandwich program, because it put a heavy burden on companies accepting engineering students. In the late 1960s, opposition to industry-academia cooperation inspired student-activism slogans and became a taboo in Japan. Strong criticism hindered the development of engineering education as vocational education in Japan. Chapter 4 discusses the details.

EARLY OPINIONS ON ENGINEERING ETHICS

Professional ethics became an issue with the promotion of vocational education at the university. Oyama Matsujiro, who became the first president of the JSEE, provided information on the American PE system to Japanese engineering professors, focusing on the engineering ethics and the training committee.

> American-style education focuses on creating respectable individual citizens, so they seem to strictly consider not only knowledge but also a higher ethical discipline for the training of engineers who must have a significant influence on society. This effort suggests what we must consider carefully when transitioning to the new university system. (Oyama 1950, 205)

The JSEE also considered the matter of "morals for engineers to preserve" at the fourth steering committee meeting in February 1953. Shimizu insisted on developing a standard code with a proposal and translated the ECPD's Faith of the Engineer in the journal of the JSEE in November 1953. However, avoiding the normative approach, the steering committee preferred a practical approach and concluded: "Because such a topic would be sensitive, it would be better to collect and publish actual cases of admirable engineers in the journal" (JSEE 1953, 100).

Engineering ethics was given attention, at least, in some industries for industry-academia cooperation. Kase Kosaku (Yawata Iron & Steel Co., Ltd.) stated: "Recently, confidentiality and other ethical issues of the advisory system and commissioned research by professors have become a popular topic." Kase provided the following examples:

> A careless word from a person who is an adviser to a similar business in several companies; employing a graduate student or assistant involved in commissioned

research at a competitor company through the recommendation of their professor (even though freedom of employment cannot be restricted, morality must be considered); misconceptions about patent rights arising from inadequate contracts for commissioned research; exploring the key point of corporate research through a former student employee; withdrawing money for commissioned research equipment without reporting it; and so on; I do not have time to give all examples. In any case, it is necessary to adopt any code of ethics or any rules for advisory and research commissions such as those in foreign countries to maintain mutual understanding and trust in these ethical issues, isn't it? This is also one of the tasks that the JSEE must consider. (Kase 1953, 60–61)

The confidentiality that Kase asked for was also emphasized in the JCEA's code of ethics adopted in 1951, which is discussed in the next chapter.

Furthermore, at the Second JSEE Annual Conference, held in July 1954, the issue of "requests and measures for engineering education from the local industry" was raised. A questionnaire survey was then conducted with twenty-one companies in various industrial fields in the Tokai[24] region. Thus, the problem of engineering ethics became clear in requests for general education and humanities education; the conclusion in the December workshop was as follows:

As for the characteristics, especially emphasized as activeness, progressiveness, devoted spirit, affection for the company, responsibility, perseverance, etc., the nature of the occupation of industrial technology and its cultural and national importance seems not to be recognized enough to be aware of their own mission. Therefore, we ask the JSEE to take measures to consider the educational method at each university and to spread it widely for students to enhance their awareness, take pride in their vocations, and engage in them with a sense of a highly cultural and national mission rather than with the passive attitude of employees. (Tokai Society for Engineering Education 1955, 139)

The statement seems to reflect prewar nationalism. The basic social perspective would not have changed easily, even when Western professional ethics was introduced after the war.

Shimizu, who became the chairperson of the Tokai Society for Engineering Education, a district of the JSEE, was most enthusiastic in formulating the JSEE's code of ethics. In the 1956 plan of the JSEE, when he became the president, the establishment of eleven committees, including the engineering ethics research committee and their assignment to each district of the JSEE, were proposed (JSEE 1956b, 236).[25] He probably modeled it on the ASEE Committee on Principles of Engineering Ethics.[26] Shimizu recalled how he prepared a regulation for the system regarding the Code of Engineering

Ethics Investigation Committee. However, it was not realized because of "the expenses, when the foundation of the organization become a little more stable" (Shimizu 1962, 13–14). After all, the ethics committee's plan dissipated while Shimizu and other representatives of the JSEE worked on industry-academia cooperation and participated in the JPC's visit to the United States.

At the end of his presidential address at the Fourth JSEE Annual Conference in May 1956, Shimizu explained the necessity of cooperation with the JPC and the Science and Technology Agency (STA) and made the following appeal:

> I sincerely hope the cooperation, which is said to be the biggest flaw of the Japanese people, will be realized in our JSEE with your enthusiasm. For this goal, I would like all engineering professors to become members of this society. I also hope that the industry will generously support this society. (Shimizu 1956, 8)

It is not clear why Shimizu described "cooperation" as the biggest flaw of the Japanese people. While changing from the prewar nationalist system to the postwar democratic system, a new democratic form of industry-academia cooperation was sought; however, it was challenging to elicit actual cooperation.[27] In improving engineering education, Japanese leaders found and appreciated American engineering ethics and emphasized the value of cooperation as the basis. The postwar emphasis on industry-academia cooperation was based on the philosophy of democracy. In electrical engineering, it was also based on the methodology of R&D management. However, in Japan, it evoked bitter memories of prewar nationalistic and authoritarian cooperation. It also evoked a feeling of rejection in many democracy advocates. Chapter 4 discusses its development into a protest movement.

In summary, this chapter discussed the relation between the engineering education reform and engineering ethics that originated in the Allied Occupation period. The next chapter discusses another aspect of this period: the relation between engineering qualification and engineering ethics.

NOTES

1. The ECL was subsequently transferred to the Nippon Telegraph and Telephone Public Corporation and is now the NTT Musashino R&D Center.

2. Yoshida (1948a, 1) stated: "At the same time that the purpose, responsibility, and organization changed, and the laboratory members' attitude was formed, the switch from academic research to industrial research, from academic shrine to developmental laboratory, from arbitrary research to organized research, from isolation to involvement with companies, from all electrical engineering to telecommunications

engineering, and so on were carried out. For us, it is a revolution of a government laboratory."

3. Although Okochi's distinction was not clear enough, he defined scientific discovery as a basis of industrialization to find a scientific theorem (causality) and described how the discovery becomes an invention by doing *jitsuyōka* for the well-being of humankind (Okochi 1935, 85). Moreover, interpreting *jitsuyōka* as a previous stage of *jissaika* (realization), he criticized the fact that it was necessary to do sufficient industrial tests for the realization of the invention, but Japan had no such facilities other than RIKEN (Okochi 1935, 83–100).

4. The English name was further changed to the current Japan Society for the Promotion of Science (JSPS) in 1963.

5. For example, the OECD defined development as "The use of the results of fundamental and applied research directed to the introduction of useful materials, devices, products, systems, and processes, or the improvement of existing ones" (OECD 1963, 12), thereby distinguishing it from fundamental and applied research.

6. See chapter 1, section 1.3.

7. Sarasohn later related that a company, which later became Sony, had manufactured high-tech devices, ordered by the Japan Broadcasting Corporation (NHK), in "shabby shacks" with leaky roofs (Hopper 1982, 15–16).

8. From September 1949, the intensive 128-hour course (for four hours a day, four days a week, and eight weeks) was held in Tokyo (1949) and Osaka (1950). Under the policy of spreading the learning from the participants' companies to those of the FJECIA and other industries, the number of participants increased from 40 in 1949 to 158 in 1950 and 1,089 in 1952. The number decreased after 1953, but the course continued until 1974.

9. Hopper arbitrary noted the influence of Zen Buddhism and Bushido, as well as that of *Self-Help* by Samuel Smiles in the Meiji era, but no historical evidence demonstrates their direct causality.

10. The former air corps and navigation school site at the First Suzuka Naval Air Station was used for the school. Tokyo Postal Service Training Center (Tokyo Yūsei Kenshūjo) and Tokyo First Telecommunications Training School (Tokyo Daiichi Denki Tsūshin Gakuen) were also established at the same time.

11. Eells had spoken of the need to exclude communist professors since July 1949 at Niigata University, which caused nationwide protests. After serving as a professor of education at Stanford University and the executive secretary of the American Association of Junior Colleges, Eells became an advisor on Higher Education at the CI&E in 1947. In general, the Allied Occupation policy initially utilized communism to achieve democratization and academic freedom. However, after the announcement of the Truman Doctrine and the Marshall Plan in 1947, an anti-Soviet and anticommunism approach became a prerequisite.

12. From February 1950 to May 1951, with the help of Polkinghorn, six engineering professors, including Koga, were dispatched to the United States by the Government Appropriation for Relief in Occupied Areas (GARIOA) fund as national leaders of Japanese engineering education. On this tour, Koga visited Harold L. Hazen (head of the Department of Electrical Engineering, MIT), who

would arrive in Japan in 1951 as the chairperson of the Engineering Education Mission.

13. By living in Japan, Polkinghorn realized that social education and engineering were closely linked: "There is no better experience than the one I have just had in coming to a foreign land with different customs and practices to impress one with the close relationship between social education and the practice of engineering. The whole approach to engineering problems is based upon social concepts which differ greatly in the two countries" (Polkinghorn 1950b, 1).

14. Although it is unclear who is referred to as Compton, the person was probably Karl T. Compton, president of MIT.

15. However, Polkinghorn did not want the university to become a vocational school. He had never argued that basic science should be reduced in engineering education (unlike training for telecommunications technicians).

16. They were corporate executives of companies such as Toshiba, Nippon Electric, Furukawa Electric, and Fuji Tsushinki Manufacturing. The CCS organized the Japanese Federation of Communications Equipment Manufacturers Association in January 1946 as an organization for controlling the telecommunications businesses. However, this association was dissolved under the Antimonopoly Act enforced in 1947. Thus, the FJECIA was launched in April 1948 as a more democratic association of 196 private companies. It expanded to about 500 companies in the 1950s. The CCS management course was decided by the board meeting of this association in August 1949.

17. He returned to the United States in May of this year.

18. The meeting was held twice at the FJECIA in March 1950 and once a month at the members' universities or companies from the third meeting. There were forty attendees from universities and twenty from the industry at the general meeting in November. The meeting continued thereafter, and the thirtieth was held in 1959 (Postwar Communications Industry Editorial Committee 1959, 243). The same meeting was also held in the Kansai area (Koga 1952, 43–44; Koga and Taki 1952b, 1).

19. They suggested that it was better to teach at every opportunity, including lectures and practical training rather than provide individual lectures focused solely on either economic or ethical issues (Koga and Taki 1952c, 2–3).

20. Hazen emphasized the importance of high ethical standards and moral fiber as safeguards against abuse, along with the professional freedom to practice engineering as a consultant (Japanese Central Executive Committee 1951, 89).

21. Shimizu became the president of Nagoya Institute of Technology in 1949, after serving as a school inspector of the MOE (1941), the principal of MCT (1944), and the director general of the MOE Science Education Bureau (1946).

22. Nanbara, who delivered a welcome speech to the Engineering Education Mission in 1951, praised the JSEE as one of the fruitful activities in the postwar education reform, and looked back on the founders' enthusiasm (Tsuzaki 1972, 2). The Mission also said that the sessions and the report were conducted "in terms of suggestions based on American experience, rather than firm recommendations" (Technical Education Division of the MOE 1952, 1). Oyama Matsujiro, dean of the Faculty of Engineering, the University of Tokyo, became the first president of the

JSEE, and Shimizu served as the executive director. The office was located in the MOE Technical Education Division.

23. Prior to this report, the twenty-first JSEE steering committee meeting in February 1956 discussed the issue, "sangaku-kyōdō seido ni tsuite" (on industry-academia cooperative system), which was also to introduce the council's activities. Before that, for example, the JSEE translated the educational system as a sangaku *kyōryoku* seido (also industry-academia cooperative system in English).

24. The region is located in the middle of Japan, between Osaka and Tokyo, including Nagoya.

25. Eleven committees were considered at the eighteenth steering committee in September 1955, including the Executive Committee for Improvement of Engineering Education (JSEE 1956a, 150). However, this committee was replaced by the Engineering Ethics Investigation Committee at the board meeting and the board of representatives in April 1956.

26. Shimizu published a history of American engineering education, including the ASEE's case of the ethics committee (Shimizu 1952, 29).

27. Japan needed to shift the cooperation from a nationalistic militant type to a democratic industrial type, based on Spencer's analysis, noted in chapter 1.

Chapter 3

Import of the Western Engineering System and Its Ethics

TWO CODES OF ETHICS ADOPTED BY THE JCEA

The preceding chapter analyzed the improvement of engineering education in Japan immediately after World War II. The CI&E of GHQ/SCAP executed the overall reform of Japan's education system as part of the Allied Occupation policy. Meanwhile, the CCS initiated the educational improvement, and Japanese engineering professors democratically drove this effort. Apart from the movement in engineering education, Japanese engineers followed the American Consulting Engineer (CE) business by attempting to establish an engineering qualification system. Engineering ethics gained popularity during this process. As discussed in the introduction, Layton generally determined that the engineers' professional system is not only autonomous and democratic but also expertly elitist, conservative, and not progressive (Layton 1971, 7–8, 65–67). The Japanese system of engineers also experienced this dilemma.

The Japan Consulting Engineers Association (JCEA, Nihon Gijutsushi-kai) adopted a code of ethics, Gijutsushi Fukumu Yōkō (the Code of Service for CEs), when the association was founded in 1951. In addition, the JCEA adopted a new code of ethics, Gijutsushi Gyōmu Rinri Yōkō (the Code of Ethics for CE Practice), ten years later in 1961. During the JCEA's first decade, the Gijutsushi Act (the CE Act, currently the PE Act) was enacted in 1957, the first JCEA was disbanded in 1958, and the JCEA was again established as a new organization that same year. The 1961 code was not *revised* but was instead newly *adopted*, which was probably because the new JCEA regarded itself as an entirely different association. Therefore, the 1961 code was based on the 1951 code's articles and expressions. However, its format and contents were substantially different.

The Creed and the Code of Practice of Civil Engineers, adopted by the JCES in 1938, is the oldest code of engineering ethics adopted by Japanese academic societies and professional engineering associations. However, the JCEA's codes are the oldest after World War II. Other Japanese engineering societies began establishing codes of ethics after the mid-1990s. Therefore, the JCEA case contributes insight into the intentions of post–World War II engineers, who needed a code of ethics, to evaluate the entire history of engineering ethics in Japan. However, no historical study analyzes the JCEA code. Even the JCEA's anniversary books do not describe the details of the 1951 code.

This chapter discusses the historical details of the 1951 and 1961 codes, problems faced by the old and new JCEA as a professional organization, and the intentions of the members responsible for resolving the problems. Comparing the two codes of ethics reveals similarities and differences in the underlying reasons for why engineers sought them in the early Japanese CE system.

HIGH-LEVEL ENGINEERS IN THE ESTABLISHMENT OF THE JCEA

It is helpful to explain the historical context of the association's establishment before analyzing JCEA's two codes of ethics. The Japanese Gijutsushi system began by establishing the JCEA in June 1951, which was modeled on the American CE profession. The Arrangement Committee for Establishing the Consulting Engineers Association was organized in December 1950, six months before the JCEA's establishment.

The main sources to analyze the JCEA's establishment process include two JCEA-edited anniversary books, which cover thirty and fifty years of history (JCEA 1981; IPEJ 2001b). Regarding the reasons the JCEA was established, both anniversary books explain that people in Japan began hoping for a Japanese CE as they became aware of the American CE's social role and status. They also explain that technology became recognized as an object of commerce since the 1950 Act on Foreign Capital had promoted the introduction of technology (JCEA 1981, 1–3).[1]

The First Yoshida Cabinet proposed the introduction of foreign private capital as a major policy in 1946, and the Third Yoshida Cabinet enacted the Act on Foreign Capital in May 1950 to remove the remaining obstacles. In this act, a technical assistance contract and the private sector's stocks and bonds were approved together. According to a memoir written by early JCEA members, CE symbolized the advanced technological power expected from introducing foreign private capital and technology in Japan. The Allied

Occupation forces aimed to use Japan's CE for civil engineering to reconstruct Japan, but they were surprised to learn that such a system did not exist in Japan. Therefore, the occupation forces sometimes commissioned the work to the construction office, which acted as an intermediate agent to the civil engineers.[2] As the introduction of technology progressed, various survey teams arrived in Japan from the United States, and American CE's participation in the survey teams began attracting Japanese engineers (JCEA 1981, 1). Moreover, Japan required consulting services for plant exports, chemical industry equipment, and construction machinery; therefore, offices and research organizations with related expertise began to be established (Tanaka 1951a, 2–3).

However, the above story is based on the recollections of the persons directly involved in the early JCEA. Therefore, there remains a need for research that examines how widely the interests and expectations of the CE system were shared in society at the time. For example, the first report of the Council for Industrial Rationalization of the MITI in February 1951 suggested the "training of consulting engineers and the dissemination of this system." However, it was only one of many suggestions (Council for Industrial Rationalization 1951, 2, 123). In addition, the council did not refer to the CE system in its second report the following year and its *White Paper on Industrial Rationalization 1957* (MITI 1957). Furthermore, there is no evidence in the minutes of the Foreign Capital Committee of the Economic Stabilization Board (Keizai Antei Honbu), which promoted the introduction of a CE system in Japan. The implementation of the Japanese CE system would have been promoted by the voluntary activities undertaken by some engineers centered on the Arrangement Committee rather than by governmental policy.[3]

Their voluntariness can be inferred from the background of the early JCEA members. Of the total 354 JCEA members in 1955, 80 were repatriated engineers from foreign lands (Korea, Taiwan, Sakhalin, Manchuria, and the South Sea Islands) and 51 were former military engineers, which led Sawai to conclude that "they were elite engineers with high educational and professional careers in the prewar and wartime periods," namely "high-level engineers." Moreover, the early JCEA served as an organization to convert military engineers into civilian ones and receive repatriated engineers from abroad during the postwar restoration period (Sawai 2012, 487–517).

Hirayama Fukujiro, who later became the third JCEA president (1959–1962), and his friend, Shiraishi Tashiro, are examples of such *high-level engineers*. Hirayama and Shiraishi were classmates at the Prefectural First Middle School, the First Higher School (the best elite school in Japan), and the Department of Civil Engineering at Tokyo Imperial University. Moreover, they were colleagues at the Railways Agency (later the Ministry of Railways). Hirayama even became a director of the South Manchuria Railway. Shiraishi became the first president of Komatsu Ltd., founded the Shiraishi Kiso Industrial Company, and was also a lecturer at Tokyo Imperial University.

The president's room in Shiraishi Kiso headquarters (Room 424 of the Marunouchi Building) became "a gathering spot for seniors and friends of Tashiro who lost their place due to the purge from public office by GHQ." Furthermore, it became "like an assembly hall for people who worried about the future of our country, which had been devastated and lost almost its production functions due to the war" (Shiraishi 1994, 164–165).[4] Inoue Tadashiro, who later became the second JCEA president (1952–1959), and Utsumi Kiyoharu, the first JCEA vice president (1951–1952), gathered in this room. Hirayama joined the group in late 1946 after returning from Manchuria.

He was also invited to Prime Minister Yoshida Shigeru's residence together with Shiraishi Tashiro and his younger brother Shiraishi Muneki[5] around this time, and Yoshida urged them to establish a CE system for postwar restoration (Kawano 1981, 14).[6] It is not clear how Yoshida knew about the CE profession or how important the establishment of a Japanese CE system was to him, as this information is only discussed in Hirayama's memoir. Yoshida might have been interested in the CE business because his nephews Shiraishi Tashiro and Muneki informed him of the profession. Shiraishi Tashiro could advance the development project of the Tadamigawa river power supply in cooperation with Erik Floor, an American CE. Later in 1949, he met directly with Yoshida and General Douglas MacArthur of GHQ/SCAP to seek their cooperation with the project.[7] These *high-level engineers*, as non-governmental private engineers, promoted the consulting business of engineers.

During this postwar outset of the engineering consulting business, Tanaka Hiroshi of the Economic Stabilization Board promoted the introduction of the CE system. From May to August 1950 (when the Act on Foreign Capital was enacted), Tanaka, as section manager of the Machinery Section of the Production Bureau,[8] traveled to the United States with Matsumura Keiichi, section manager of the Industrial Policy Section, to inspect the country's industrial administration, industrial association, and association of engineers. Tanaka also surveyed the CE business in the United States and found that it ensured "a proper reward for their engineering work, while moral arrangements for the profession are well adhered to between the engineers" (Tanaka 1951a, 30). He also found that the code of ethics supported the social status of PEs. After surveying the American system, he understood that to develop a CE system in Japan, it was necessary "to establish regulations concerning qualifications and register eligible engineers as a means of strengthening the notion of responsibility of consulting engineers and improving their ability." Moreover, "it is also necessary to agree on the code of ethics because the professional moral is an issue particularly in this occupation" and engineers "should strive to improve their qualitative contents" (Tanaka 1951c, 5).

Tanaka also found no official CE qualification system but a more comprehensive PE license system in the United States. The PE system was examined and registered by each state, and it also functioned as a public qualification system for CE services (Tanaka 1951a, 12–22). Tanaka aimed to introduce the American CE profession to Japan. Thus, he returned to Japan and reported on his surveying results. He received "unexpected great responses" (Tanaka 1950a, 45) from interested parties.[9] Thus, in December of the same year, the Arrangement Committee for the Establishment of the Consulting Engineers Association was organized to introduce the CE profession to Japan. This committee consisted of eight engineers: Asahara Genshichi, Hayasaka Chikara, Kato Mieji, Ohno Iwao, Tanaka Hiroshi, Toriya Torao, Yagi Susumu, and Yoshimura Masamitsu. The committee members belonged to various technical disciplines, such as mechanical, metals, civil engineering, industrial engineering, and applied science.

Asahara, a member of the Arrangement Committee, reported: "On December 14, 1950, concerned volunteers gathered on the Economic Stabilization Board's initiative [. . .], and they exchanged their opinions on the creation of a new occupation called Consulting Engineer in Japan" (Asahara 1966, 12). With this statement, Asahara noted that *concerned volunteers* conducted this arrangement activity. Although Asahara referred to this activity as the Economic Stabilization Board's initiative, it was Tanaka's initiative, not the board's policy, as reflected in Tanaka's proposal to the government that "it is appropriate for the government to assist and cooperate in the right direction, as a means of encouraging our nation's development of industrial technology" (Tanaka 1950a, 46).

THE 1951 CODE OF ETHICS OF THE JCEA

The Arrangement Committee defined the name "Gijutsushi" and drafted its articles of association, including the Hōshū Kitei (Terms of Compensation), Shikaku Kitei (Terms of Qualification), and Fukumu Kitei (Terms of Service, later renamed "Fukumu Yōkō" [the Code of Service]). Finally, the Gijutsushi Fukumu Yōkō was adopted when the JCEA was founded at its inaugural meeting in June 1951 (Takada 1958, 12–13). Although the title did not include the word *rinri* (ethics), this code of service was adopted as a code of ethics. In its preparatory stage in 1951, Tanaka reported that the Arrangement Committee was "steadily creating the regulations of the association, fees, qualifications, ethics, etc." (Tanaka 1951a, 43). Asahara, an Arrangement Committee member, also stated that the Fukumu Yōkō's professed moral set the basic conditions to socially establish the Japanese Gijutsushi system as follows:

> I remember that the Gijutsushi Fukumu Yōkō was the rule that was considered to be the most important among the participants, which was subjected to the solemn discussion. Although the charter on the morals of Gijutsushi, which was made in the founding period over fifteen years ago, was drafted by being modeled on the codes of advanced countries at that time, it was not merely an imitation of the format but also a declaration of the basic conditions. Aside from the skillfulness of the phraseology, to determine whether an occupation called Gijutsushi could subsist in society by examining whether the spirit shown in the code was followed or not, neither *Gijutsushi* nor the *JCEA* will be able to subsist without their sufficient understanding and implementation. (Asahara 1966, 13–14)

Asahara and Tanaka both recognized that the code of ethics contributed to the CE's social status.

Hirayama specifically stated that this code was created "by referring to the codes of ethics of some American engineering associations and the consulting engineers' association" (Shiota 1951, 49). Based on Tanaka's U.S. report and other documents, the associations that Hirayama referred to are presumably the ECPD and the AICE (Tanaka 1951a, 29–36).

The AICE's code of ethics was the most informative reference used in preparing the JCEA code. In his U.S. report, Tanaka translated the AICE code's full text. More so, in the handout prepared for the 1960 JCEA board meeting, the "MEMO on the Gijutsushi Fukumu Yōkō," the JCEA and AICE codes were mentioned with respect to their sections of clear contract, unfair competition, authority of expertise, and cooperation with other professional engineers. For example, in the 1960 memo, Article 3 in the 1951 JCEA code was transcribed into the *clear contract* item and annotated with Gi3. The translated version of Articles 1, 2, and 4 in the AICE code, which was published in Tanaka's 1951 U.S. report, were also transcribed into this section and annotated with A1, A2, and A4, respectively (JCEA 1961a). In addition, the Canons of Ethics for Engineers, adopted by the ECPD in 1947, became a model code of ethics for various engineering associations, such as NSPE (*Chemical & Engineering News* 1948; Baum 1980, 8–9; Davis 1998, 45–47). Tanaka noted that it was the "most authoritative" code in the United States (Tanaka 1951a, 31). Regarding the *clear contract* section, the Japanese-translated ECPD code articles were transcribed into the memo and annotated as EC9, EC16, and EC18; thus, it appears that these ECPD articles were used to develop the JCEA code's Article 3 (JCEA 1961a).

In addition, the JCEA code included Articles 1 (maintenance of dignity) and 2 (maintenance of confidentiality). The AICE code refers to dignity in

Article 11 (restriction on advertising), and Tanaka's translation in his U.S. report acknowledged that a CE should not "publicly advertise his professional practice, qualifications, or achievements, in a misleading, self-laudatory, or undignified language or manner" (Tanaka 1951a, 32, 35–36). However, in the 1951 JCEA code, dignity was disassociated from the restriction on advertising and was instead generalized as a *cultivation of dignity*. Regarding this "dignity," Hirayama discussed the professional ethics of Japanese engineers in 1951 as follows:

> Next, on the ethics issue of professional engineers, the sense of engineering as a profession is poor in the Japanese engineering community, and the system is not legally established. So, they generally have little awareness of and interest in the engineering profession. In contrast, the American engineering community always discusses and strives to improve the engineering profession's virtue, authority, and dignity as a problem of engineers. Generally speaking, although a sense of professional ethics is rare in Japan, this is a problem that must be considered for the development of the profession. (Shiota 1951, 49)

Unlike in the United States, it was rare for Japanese engineers to uphold professional ethics. Therefore, the maintenance of dignity was regarded not as the specific advertising restriction problem but a general problem that would be encountered when introducing the CE system into Japan.

Furthermore, Japan uniquely introduced its article on the maintenance of confidentiality, as neither the AICE nor ECPD codes of ethics contained related sections. In his U.S. report, Tanaka (1951a, 1–2, 42) mentioned that clients' suspicions that confidentiality was not being maintained were a major reason the CE system was not popular in Japan. Moreover, Tanaka (1950b, 19; 1951b, 17) explained that, in the case of the United States, confidentiality was no longer an issue and was not specified in the code of ethics because U.S. engineering contracts sufficiently covered this topic. Maintenance of confidentiality thus had to be explicitly stated in the Japanese code of ethics to improve clients' trust in engineers. Their high regard for confidentiality can be observed in that only the phrase "the secret must be kept absolutely" was emphasized with a gothic bold font style in the JCEA's journal publication of its code of ethics (JCEA 1951b, 8).

The maintenance of dignity and confidentiality was particularly emphasized in Japan. Thus, both elements were listed as the first and second articles of the JCEA's code of ethics in 1951. Meanwhile, the 1951 code did not refer to "independent of all commercial interest," which later became a serious issue that compelled the JCEA to adopt a new code in 1961—the Gijutsushi Gyōmu Rinri Yōkō.

THE 1961 CODE OF ETHICS OF THE JCEA

At the JCEA's inaugural meeting, the secretary of the Agency for Industrial Science and Technology and the deputies of GHQ/SCAP, the prime minister, the director of the Economic Stabilization Board, and the MITI all sent "greetings of encouragement" (JCEA 1951a, 1; Shiota 1951, 47; Asahara 1966, 12). Thus, both the Japanese and U.S. governments generally welcomed the new Japanese CE system. However, this "encouragement" indicated that the initiative and responsibility for this new system had been entirely entrusted to JCEA members. Therefore, immediately after its establishment, the JCEA (particularly its board members) painstakingly promoted the CE system's utilization to achieve the desired status of engineers. The JCEA also conducted the preparation of the 1961 code to meet this purpose.

It hoped that the enactment of the Japanese CE Act in 1957 would greatly encourage society's use of the Japanese CE system. However, the reality was "disappointing" (Murakawa 1981, 90). Therefore, in August 1959, the JCEA President Hirayama submitted "A Request for the Utilization of Gijutsushi" to the STA secretary to improve the situation.[10] In the fall of 1959, in response to this request, the STA asked the JCEA to survey the actual conditions of the European and American CE systems (JCEA 1960d, 3). Japanese government offices abroad first surveyed the European CE systems in cooperation with the Ministry of Foreign Affairs and learned about the International Federation of Consulting Engineers (Fédération Internationale des Ingénieurs-Conseils, FIDIC).[11]

FIDIC was founded mainly by the Belgian and French CE association members at the World Fair Exposition in Ghent, Belgium, in 1913. The original aim was only to secure a certain CE's post to the world expo jury members, but it developed into a federation of the European CE associations due to its benefits in promoting technology trade. Although the scale was temporarily reduced during World War II, the United States, Canada, South Africa, and other non-European countries consecutively joined in 1951, and FIDIC developed into a worldwide federation (Widegren 1988, 33–36, 138). Two years prior, in March 1957, the European Economic Community was signed at approximately the same time Japanese officials began to survey the European CE systems.

This preliminary survey clarified that FIDIC strictly set the code of ethics as a CE and demanded that all members follow it. In particular, FIDIC's code of ethics requested independence and neutrality as in any noble profession (e.g., lawyer and medical doctor). Specifically, it prohibited engineers from obtaining compensation from anyone other than the client, and it requested that engineers be free from commercial biases (Widegren 1988, 124–125). However, many JCEA members managed related businesses or became members of related companies. Thus, they did not earn their income independently or only from their consulting business. Therefore, the JCEA was initially denied into FIDIC's general assembly even as an observer because FIDIC's demand for members to be "independent of all commercial interest" was strong (JCEA 1981, 82–85). However, after negotiations, two JCEA members were eventually allowed to attend the general assembly in June 1960, and six JCEA members visited several European countries to survey each CE association and system.

This problem arose originally from the unique definition of the Japanese CE system, which was introduced by the Japanese CE Act in 1957 and differed from the original CE system in other countries. At first, the 1954 draft of the CE Act clearly stated in Article 2 that Gijutsushi could "earn rewards according to the request of others." As Hirayama (1961, 134–136) stated, the premise was to do consulting business "according to the request (order) of clients (customers)." If this premise was eliminated, civil servants and corporate employees who were "employed by their clients" could also be included in the Japanese CE. However, this requirement was removed in the final version that was enacted in 1957. The business of Gijutsushi was extended to include general *high-level engineers*, as evidenced by Hirayama's statement: "the qualifying examination of Gijutsushi is a national certifying examination of high-level engineers" (Hirayama 1960, 2–3). Given this unique Japanese CE definition beyond the original restriction, it became inherently impossible for the Japanese system to meet FIDIC's criteria of being *independent of all commercial interest*.

Furthermore, the Japanese CE Act only licensed the Gijutsushi name and did not provide occupational licensing. The JCEA, therefore, aimed to amend the CE Act to acquire occupational licensing for the Japanese CE to improve engineers' social status. In fact, JCEA President Hirayama (1960, 3) wrote that "high-level engineers must dare to become Gijutsushi, dare to engage in social activities as consulting engineers voluntarily, and dare to achieve technical authority and social trust of the Gijutsushi system." Hirayama (1960, 3–4) also argued that two issues must be addressed for the CE system to gain authority and trust: "the problem of the nature of Gijutsushi" and "the problem of occupational licensing." Japan's CE problem was that their independence and neutrality were ambiguous, which became explicit by inquiring about FIDIC. Thus, considering these problems, the need to amend both the JCEA's code of ethics and the Japanese CE Act were discussed in parallel.

First, at the October 1960 JCEA board meeting, "the code of ethics for Gijutsushi" was proposed and scheduled for the next meeting's agenda (JCEA 1960b).[12] Moreover, the 1951 JCEA code, the Rules of the Japan Association of Patent Attorneys, the Rules of the Japan Federation of Bar Associations, the Rules of the Japanese Institute of Certified Public Accountants, Charter of the Japan Institute of Architects, and "the European code of ethics (the Netherlands Consulting Engineers Association)" were all circulated as references. It is important to note that the European code was referenced in this meeting, unlike in 1951, when the American code was referenced. The Netherlands' CE association, the Orde van Nederlandse Raadgevende Ingenieurs (ONRI), replied to the preliminary survey on the European CE systems. The handout of the European code was a direct translation of the ONRI's code (JCEA 1961d, 13), which was largely characterized by three requirements: independence and neutrality, authority of and cooperation with CE, and restrictions on advertising. A handwritten note was then added to specify that the code was "much the same as in Europe." The independence and neutrality explanations specified that "FIDIC Article 3 (6), there is an article with the same meaning." Thus, the JCEA relied on the ONRI and FIDIC to develop the code's requirements.

At the ninth board meeting in November 1960, the JCEA decided to draft the code of ethics simultaneously with the proposed CE Act amendments (JCEA 1960c),[13] and members circulated the "MEMO on the Gijutsushi Fukumu Yōkō." At this meeting, they discussed the necessity of including maintenance of dignity, maintenance of confidentiality, unfair competition, independence and neutrality of the position, authority of expertise, restriction on advertising, and cooperation with other professional engineers in the code of ethics. The items, *independence and neutrality of the position* and *restriction on advertising*, had not previously been included in the 1951 code and were remarked in the ONRI's reply. The memo relied on the ONRI code and the Rules of the Japan Association of Patent Attorneys (Articles 24 and 25) to provide examples of these items (JCEA 1960c).

This memo was circulated at the eleventh board meeting in January 1961, and the items, *clear contract* and *report to clients*, were added to the memo, with further detailed analysis on the code as a whole. At this meeting, the JCEA decided that there would be a special committee on the code of ethics, and Tanaka would chair the committee. In the revised memo, the Association of Consulting Engineers (ACE) in the United Kingdom, the ONRI, and FIDIC were referenced as the European codes of ethics. The American AICE and ECPD codes, which had been originally included in the 1951 code, were also referenced (JCEA 1961a).[14] The memo referenced numerous American AICE's code items but included only one AICE item on the issue of independence and neutrality. In contrast, the memo included five items on independence and neutrality from European associations (three from the ACE, one from FIDIC, and one from the ONRI). This fact demonstrates that the JCEA largely developed its independence and neutrality items to comply with the European standard. JCEA members continued to discuss the code of ethics at the twelfth board meeting in February 1961 and presented a revised draft. A report stated: "Although various opinions were stated, our discussions were focused on the article on *neutrality of the position*" (JCEA 1961b).

It is unclear whether JCEA members originally failed to recognize the significance of *neutrality of the position* or if they intentionally dismissed it because it seemed too difficult for members to accomplish this goal. The AICE code of ethics, used to prepare the 1951 JCEA code, included an article (Article 12) that prohibited engineers from "submit[ing] proposals for or enter[ing] into contracts for the construction of works, plans and specifications for which have been prepared by him in the capacity of professional Engineer for a client." Moreover, the article independently distinguishes any consulting business. However, the JCEA failed to recognize the article's standing for the importance of the position's neutrality in 1951 as it was not included until the January 1961 handout. In his 1951 report, which included a summary of the AICE code of ethics, Tanaka failed to mention neutrality (Tanaka 1951c, 4).

The JCEA, thus, adopted the new code of ethics at the thirteenth board meeting on March 14, 1961 (JCEA 1961c).[15] The name was changed from "Gijutsushi Fukumu Yōkō" (the Code of Service for CEs) to the "Gijutsushi Gyōmu Rinri Yōkō" (the Code of Ethics for CE Practice) at the meeting. The underlying reason for this change is unknown, but the JCEA likely wanted to emphasize that the code included "ethics" and stress that the code belonged to an entirely new organization. The old name was used until 1961, as evidenced by the fact that the handout was named the "MEMO on the Gijutsushi Fukumu Yōkō" and the final February version was named "Gijutsushi Fukumu Yōkō." However, the old name and other names such as "Gijutsushi Rinri Kitei" (the Code of Ethics for the CEs), "Rinri Kitei," and "Gijutsushi-kai Fukumu Kitei" (the JCEA Code of Service) were included in the board meeting's minutes. Therefore, the JCEA conducted this work to establish a new code for the new JCEA rather than revise the old 1951 code.

In the new code, Article 3 stipulated the importance of the neutrality of the position: "Gijutsushi shall not manage nor be employed in any other business related to his or her own Gijutsushi business (except when it mainly consists of Gijutsushi business)." A transitional measure was included as a supplementary provision, which stated, "As for the neutrality of the position, regardless of Article 3, the next interim measure shall be admitted for a while." Regarding the problem of neutrality of the position, Hirayama stated the following in 1960:

> For this reason, I think that the Japanese Gijutsushi has come to a stage where they should emphasize the neutrality problem; but even if it is impossible to

implement them as soon as in the advanced countries, at least the Gijutsushi must stop serving concurrently as executive or employee of a commercial company and so on, by all means. I have not settled this relationship yet, so it is presumptuous to say such a thing, but I am convinced that it is time for this neutrality to be realized. (Hirayama 1960, 4)

While neutrality should have been an inevitable requirement in the CE's philosophy, it was challenging for the Japanese CEs to clarify it in the code of ethics and conduct it without exceptions, as in the case of President Hirayama.

Solving the commercial independence problem is challenging because the Japanese CE Act was not designed for consulting services, and the JCEA was not a CE association (Hirayama 1960, 4). In 1974, more than ten years after the revision of the code of ethics, FIDIC approved the Association of Japanese Consulting Engineers' (AJCE) admission. The AJCE was derived from the JCEA. The FIDIC's seventy-fifth-anniversary book referred with interest that Kawano Yasuo,[16] chairperson of the AJCE, regarded the government, private enterprise, and organized labor's unique "joining of forces" as crucial in Japan, illustrating that the Gijutsushi was a PE, not a CE. The FIDIC book also stated that the English translation of Gijutsushi-kai, "the Japan Consulting Engineers Association," was misleading (Widegren 1988, 192). This underlying problem of the JCEA regarding the difference between CE and PE developed into a critical issue in the 1990s, along with the additional revision problem.

CODE OF ETHICS FOR ENGINEERS' SOCIAL STATUS

Inoue Tadashiro and Hirayama attended the 1953 House of Representatives' Special Committee on Science and Technology Promotion Measures as witnesses for later deliberations on the Japanese CE Act. According to Hirayama, Japanese society tends to have lower opinions of Japanese technologies and engineers (relative to their Western counterparts). Consequently, a Gijutsushi-like system took a long time to progress in Japan. Thus, he maintained the importance of enacting the new Japanese CE Act to establish a better social status for Japan's "top engineers" as follows:

In Japan, there is no social status for top engineers, unlike doctors and lawyers, and the success of an engineer is not to be a top engineer but to be a vice-secretary, a secretary, a president, an executive, and so on. Although it may be somewhat inappropriate to say to throw technology away, this is a wrong way from the viewpoint of the promotion of technology, and I think that it is needless to say that it is a truly regrettable point. (Special Committee on Science and Technology Promotion Measures 19 (26th Diet Rec. Mar. 26, 1957))

The Japanese CE system was institutionalized to meet the unique requirements in Japan while simultaneously following the American engineering system. Therefore, the Japanese system was fundamentally designed for consulting businesses; however, it was expected that the system would improve the status of high-level engineers. While it became a deficiency of the Japanese CE Act after 1957 regarding the occupational monopoly, engineers had actively promoted it in the first place. The CE Act only licensed the Gijutsushi name and did not provide occupational licensing.

The elite high-level engineers (i.e., engineers who conducted comprehensive design work, such as consultants) found their potential as engineers in Japan's new CE system when they sought prospects for postwar reconstruction. Thus, they actively and autonomously worked to establish the important issue of their social status in this system. Tanaka appreciated that the Agency for Industrial Science and Technology of the MITI (the competent ministry) provided the office room, and the agency's secretary provided governmental cooperation in establishing the Japanese CE Act after the Economic Stabilization Board was abolished (Utsumi et al. 1966, 32–33). The fact that he was grateful for the governmental support, in combination with the fact that the CE Act was once abolished in 1954 due to ministerial objections about demand and jurisdiction,[17] demonstrates that it was the engineers (rather than the government) who independently promoted the Japanese CE system's social status.

In this situation, the engineers actively encouraged society to recognize the new system. As discussed above, the Japanese code of ethics was established based on the American CE model, which had a high status. However, even after the Japanese CE system was established in 1951, the system's utilization and popularization did not progress as expected. Moreover, its social status remained problematic. Thus, to achieve these goals by complying with the FIDIC standard, it became necessary to revise the code of ethics and declare the Japanese CE's neutrality. When the Gyōmu Rinri Yōkō was adopted in 1961, Hirayama (1961, 140) stated: "In Japan, the interest in this professional ethics is generally low, but this is particularly important for a new occupation to be developed."

It was similar to the case of the prewar JCES that the JCEA was aware that the establishment of a code of ethics leads the professional association to achieve a higher social status. The Creed and the Code of Practice of the JCES, discussed in chapter 1, was the oldest code of ethics in Japanese engineering societies. Hirayama and Utsumi (they later led the JCEA) attended the prewar JCES' investigation committee meetings, which prepared the code of ethics in the late 1930s. Hirayama attended the first and the second meetings as a general manager, Miyamoto Takenosuke was replaced after the third meeting, and Utsumi attended only the fourth meeting (Furuki and Sakamoto 2004, 71–73). Given this pattern, they likely did not contribute to the JCES code, which likely did not influence the JCEA code, because its purpose and contents were different. The establishment of the JCEA code was conducted independently as its members aimed to develop a postwar engineering system.

The above relationship between the establishment of the code of ethics and the professional association achieving social status can also be observed after the JCEA adopted the 1961 code. In July 1961, several months after the code was adopted, a survey of the American CE system was conducted. The survey had been requested in 1959, as mentioned above, along with the European systems. In this report, the survey committee reiterated that they were especially impressed with their concept of the profession, the accompanying ethical responsibility, and their high awareness. They also stressed the importance of the code of ethics that specified their ethical standard (Gijutsu Senmonka Katsudo Shisatsudan 1963). The survey team's leader, Fukuda Eikichi, summarized the survey committee's opinions: "We, the JCEA, would like to sum up the rules of members and to revise the code of ethics by all means" (Gijutsu Senmonka Katsudo Shisatsudan 1963, 83).

The insufficiency of the new 1961 code was recognized, but a further revision of the code (including its supplements) was never progressed. Although neutrality became a problem in the 1970s regarding FIDIC admission, establishing the AJCE resolved the problem. The JCEA, however, did not address within-organization ethical issues until the mid-1990s. The ethics committee was established in 1970 to commend the members and maintain their dignity (JCEA 1970). Thus, this committee was responsible for expelling members who detracted the order and trust of the JCEA and the dignity of Gijutsushi, but such a negative issue did not occur until 1999 (IPEJ 2001b, 88–90).

However, the JCEA members' personal interest in engineering ethics is another issue. For example, Honda Naoshi, a JCEA board member, emphasized the need for Gijutsushi's ethics in his 1989 book, *Gijutsushi eno izanai* (Invitation to Gijutsushi). However, he argued that ethics should be established at the individual engineer level rather than at an organizational level to better create society-wide happiness[18] (Honda 1989, 94–98). Therefore, such an opinion did not engender any change to the JCEA as an organization. This situation remained because developing the code of ethics was only one of the necessary conditions for establishing the Japanese CEs' social status. Moreover, the association's ethical goals were temporarily fulfilled (apart from achieving social status) when the code was established per the advanced Western countries' standard of CE.

BETWEEN DEMOCRACY AND ELITISM

This chapter has discussed how the JCEA adopted two codes of ethics: the Gijutsushi Fukumu Yōkō (the Code of Service for CEs) in 1951 and the Gijutsushi Gyōmu Rinri Yōkō (the Code of Ethics for CE Practice) in 1961. These two codes had similar purposes, as they both attempted to clarify the CE business standard in the Japanese context to establish engineers' status. Both codes followed similar establishment processes, as they were both voluntarily and autonomously promoted by the engineers. While Japan simultaneously introduced foreign private capital and technology, high-level engineers (rather than the government or GHQ/SCAP policy) actively promoted the establishment of the JCEA to develop their own system and promote its status. Inoue Tadashiro discussed the Japanese CE system to the Special Committee of the House of Representatives in 1953 as follows:

> Although I think that such a system must exist, especially in Japan, considering why such a thing has not come into existence up to now, the general society and the people have remarkably lacked the idea of respecting technology. I think that this is a remnant of the feudal system thinking after all. In addition, engineers have also regarded technology as the only tool for managers' use until now and have not had the courage to confirm its independence exactly. This situation has been on going until today. (Special Committee on Science and Technology Promotion Measures 19 (26th Diet Rec. Mar. 26, 1957))

Inoue referred to the *zaibatsu* conglomerate system in prewar Japan as a factor that prevented engineers' independence and stated that the system discouraged the consultation of outsiders about the company's inside technology.

While Inoue's historical understanding followed that of the Allied Occupation policy, he thought that the engineers' system could only be developed by ridding the country of its prewar feudalistic thought and replacing it with the new, postwar democratic ideas. Yoshimura Masamitsu, an Arrangement Committee member, reported a similar understanding:

> Because this work uses technology at the request of others, it cannot be established without a democratic social system. A division of work improves the efficiency, rationalizing the management. Therefore, in countries with democratic industrial structures, research and engineering have been separated from production companies and made independent, thereby progressing as a system of consulting engineers. Our country opened the door for the industry from the beginning of the Meiji era, but due to the feudality of engineers, the industrial technologies did not appreciate cooperation and went along with the *sakoku* [national isolation] principle for a long time. (Yoshimura 1952, 11)

Thus, the Japanese engineers' system was developed based on postwar democracy. It was understood that the code of ethics was a key requirement of the new system, which was modeled on the American codes but developed based on the JCEA's then-current problem. This approach was not only democratic but also elitist and conservative. Therefore, it is questionable how far the JCEA and its codes of ethics actually embody democratic values.

On the one hand, the old and new codes of ethics had basic similarities. On the other hand, each code introduced different viewpoints via facing different problems in the varying development stages of the Japanese CE system. The first 1951 code clarified the association's policy on their concerns to popularize the completely new business style of the Japanese CE. While the 1951 code's format and contents followed the American precedents, its emphasis on maintenance of dignity and confidentiality was unique to the Japanese code and addressed distinctive Japanese CE system problems. Due to the enactment of the Japanese CE Act, the JCEA established the second 1961 code as a new organization. In addition, a major goal of the new code was to meet the FIDIC's international (particularly European) standard to increase the utilization and popularization of the Japanese CE system. Therefore, the new code addressed the issue of *independence and neutrality of the position*, which was not explicitly addressed in the 1951 code. This independence and neutrality issue stemmed from Japan's unique history that the Gijutsushi system was originally intended to establish the social status of Japanese engineers (especially high-level engineers) and not to create a strict system of CEs. Therefore, the system was initially designed to meet only the former expectation, leading to a lack of independence and neutrality in the first code.

The problems of establishing the JCEA's codes of ethics were resolved by following the American and European exemplars. However, economic globalization in the mid-1990s caused another problem, which the JCEA voluntarily and autonomously also overcame. Chapter 6 discusses the problem in the 1990s.

NOTES

1. Moreover, the JCEA's fiftieth-anniversary book suggests that the prewar engineers tended to consider technology as a mere management tool. Furthermore, these engineers assumed that this consideration encouraged the "reckless war" (IPEJ 2001b, 12–14).

2. This information is from the memoir of Hiki Hajime (civil engineering discipline). Hiki served as a JCEA board member and later as its president (Utsumi et al. 1966, 32).

3. The JCEA's fiftieth-anniversary book praised the voluntary activities of the engineers involved in the preparation of the JCEA's establishment and explained that "the background of the establishment of the JCEA is to emerge naturally by refining the history and tracing the enthusiasm and the motive power of competent people who actively participated" (IPEJ 2001b, 12).

4. Masago Juzo, executive director of Shiraishi Kiso, resigned from the company and became a consultant to the Civil Transportation Section of GHQ/SCAP as a temporary staff member of the Ministry of Home Affairs in 1947. Testimony indicates that this served as a "private diplomatic" route between Shiraishi Tashiro and GHQ (Shiraishi 1994, 163).

5. Shiraishi Muneki specialized in electrical engineering and was in charge of the construction of the Bujeon-Gang hydroelectric plant as the manager of the Heungnam factory of the Korea Nitrogen Fertilizer Co., Ltd. (Chōsen Chisso Hiryō).

6. Yoshida reportedly noted that the absence of a CE system led Japan to the reckless war and defeat. The date of this visit is unclear, but it probably took place in December 1946 (Pacific Consultants 2002, 6; IPEJ 2001b, 14).

7. The correspondence between Shiraishi Tashiro and his friend Antonin Raymond, an architect, resumed after a wartime break. By Raymond's introduction, Erik Floor, a CE from the United States, was asked to cooperate with the Tadamigawa development project. This opportunity created a special connection between them and American CE. To promote this project, Shiraishi and Raymond met directly with Yoshida and General MacArthur to ask for their cooperation. In 1949, at the same time that the project was underway, the name "Tuesday Group of Engineering Consultants" was given to the Room 424 meetings. Pacific Consultants Inc. was established in the United States in 1951 and Pacific Consultants Co., Ltd. was established in Japan in 1954, both based on the Tuesday Group.

8. Due to organizational reform, the Production Bureau became the Industry Bureau in June 1950. Tanaka became the section manager of the Technology Section of the Industry Bureau in September 1950.

9. Matsumura (1950, 56–57) also reported that the information on CE and management consultants attracted "the greatest response in the content of my survey."

10. The STA widely distributed this request letter, "On the Utilization of Gijutsushi," in November 1961 to the respective secretaries and directors of ministries and public corporations.

11. Although the JCEA journal introduced FIDIC in 1955, the introduction was only a reference and did not refer to the problems with accessing FIDIC, such as independence of all commercial interest (JCEA 1955, 6–9).

12. At this seventh meeting, "the code of ethics for Gijutsushi" was listed as Agenda 4, and the other agenda items included the CE Act amendment (Agenda 2) and the STA qualifying examination (Agenda 3). The code of ethics was listed above the JCEA survey team's report on Europe (Agenda 7). At the sixth board meeting, they agreed to hear the European report at the next meeting. These were related topics, but the code of ethics was too important to wait for the European report to be discussed.

13. The Gijutsushi Act Research Committee was organized in June 1962 to amend the CE Act, and the board adopted the final draft in October. In the proposed revision of the Act, *neutrality of Gijutsushi* and *restriction on advertising* were included in Article 24 (prohibition of dishonorable conduct). This policy was consistent with the 1961 code of ethics.

14. The British CE association was probably referenced here for some reasons: the survey highlighted its code; the United Kingdom was seen as an original consulting business country; and Julian Tritton, president of FIDIC, was a British CE member (JCEA 1960d, 1–34).

15. The adoption date was recorded as May 27, 1960, in the chronological table in the JCEA's anniversary books. However, there is no evidence of this in the board meeting's minutes. The regular general meeting was held on May 25, 1960, but there is no report on such a code of ethics at this meeting (JCEA 1960a). The European survey team departed on May 27; thus, it is possible that they intentionally adjusted the date to make it in time. However, it is also possible that there was a mistake.

16. Kawano was involved in the foundation of Pacific Consultants Co., Ltd. Later, he became its president and was also important in establishing the AJCE to join FIDIC.

17. For instance, to cite a governmental opinion, the cabinet secretary once questioned the need for this legislation due to a lack in strong demand for consulting engineers in Japan relative to the United States (Cabinet Secretary 1953).

18. As mentioned in his book, this recognition was influenced by corporate scandals that were occurring at that time, such as Toshiba Machine–Kongsberg's CoCom incident in 1987.

Chapter 4

Industry-Academia Cooperation
The Ideal and the Real

THE TABOO OF INDUSTRY-ACADEMIA COOPERATION

As described in chapter 2, leaders of the engineering education reform under the Allied Occupation were concerned with engineering ethics and considered *cooperation* the most important value. Thus, to realize this cooperation between industry and academia, they established the JSEE. It was modeled on the *cooperative system* or *sandwich program* of engineering education in the United States, which originated with Herman Schneider at the University of Cincinnati. This emphasis on industry-academia cooperation can naturally be accepted as a positive attribute in the post–World War II democratization when both sides are firmly independent, and the contents are desirable. In Japan, however, there was extreme criticism of academic independence (*gakumon no shutaisei*) being lost through cooperation with industry, partly out of fear of returning to a prewar militaristic regime. Furthermore, the industry's expectations of cooperation were relatively weak. Thus, for about thirty years, industry-academia cooperation was negatively regarded as *sangaku-kyōdō* in Japanese society, even a "taboo." This social sentiment further separated engineering education from vocational or professional education, widening the gap between universities and industry.

Inose Hiroshi, an engineering professor at the University of Tokyo, noted the strong antiwar sentiment of the 1960s student activism (particularly among social and natural sciences faculties), directed against prewar military-industry-academia cooperation (Inose et al. 1982). In 1950, the SCJ implemented the Declaration of Resolve Never to Engage in Scientific Research for the Purpose of War. In the mid-1960s, as the anti–Vietnam War movement expanded globally, criticism of military-academia cooperation gained traction, and the Physical Society of Japan adopted Resolution Three, which stated that "the Physical Society of Japan will never have any assistance from or any other cooperative relationship with any military in the future, whether domestic or foreign." While antimilitarism explains some of the antipathies toward industry-academia cooperation, such cooperation was also condemned from the academic independence perspective. Thus, as a basic policy in the Confirmation with the Delegates of Seven Faculties for the student protest against the University of Tokyo in 1969, the council of the university affirmed: "We should deny *industry-academia cooperation* if research at the university loses its autonomy to serve the interest of capital" (Kato 1969, 7–12).

Thus, the criticism of industry-academia cooperation spread throughout the 1960s, with various opinions offered. The University of Tokyo's denial of industry-academia cooperation, for example, was excluded from the final version of the Confirmation because the Faculty of Science did not sign the draft, and it was downgraded to a supplement to the basic stance of the council.[1] As discussed in the next chapter, due to the taboo nature of a debate on industry-academia cooperation various opinions on both sides were insufficiently examined, which allowed the industrial policy to promote industry-academia cooperation and engineering education reform in the 1980s. This chapter focuses on industry, student, and engineering professor interests at each historical stage to provide a multilateral perspective on cooperation from the 1950s to 1960s when the idea of industry-academia cooperation was altered and even anathematized. Although it scarcely discusses engineering ethics, this problem became an issue in the engineering education reforms since the 1980s, which eventually promoted the introduction of engineering ethics. Therefore, this chapter is essential for understanding the history of engineering ethics in Japan.

COOPERATION AS A SOLUTION TO THE SHORTFALL OF ENGINEERS

In chapter 2, the idea of industry-academia cooperation expanded into an abstract concept as part of the national industrial policy. Despite the reluctance to supporting the cooperative education system requested by engineering professors, the industry actively appealed for a solution to the shortfall of engineers due to rapid economic growth. This industrial self-interest resulted in an alteration of the idea of industry-academia cooperation that engineering professors anticipated.

In November 1956, Nikkeiren (the Japan Federation of Employers' Associations) published a proposal titled *Opinions on Technology Education*

Suitable for the Demands of the New Era, which forecasted engineer and technician numbers and requested the introduction of five-year professional colleges by merging junior colleges and high schools. The plan was to curtail the law and humanities departments, expand the science and engineering departments, and enrich the contents of specialized subjects, including engineering ethics, to train engineers and technicians in response to future economic development and improve education for science and engineering colleges for training engineers (Nikkeiren 1956). Nikkeiren complained that the overemphasis on law and humanities at universities was contrary to industry demands because the law-humanities to science-engineering student ratio was 72:25, relative to the prewar level of 65:35. Therefore, in 1957, Nikkeiren published another report titled *Opinions on the Promotion of Science and Technology Education*, a revision to the 1956 opinion.

Given this industry request, the Central Council for Education (1957) of the MOE submitted a report titled *On the Promotion Measures for Science and Technology Education* and proposed "establishing a definite industrial development policy, making an annual plan to cultivate scientists and engineers based on this policy, and taking measures necessary to carry it out." They also planned to increase faculties and departments, thereby increasing student capacity, and requested that closer universities-industry contacts be established to promote mutual cooperation "by reflecting the demands of the industry in university education and research, and asking for the industry's cooperation for students' factory practice."

In 1957, it was predicted that, by 1962, the demand for engineers would reach 27,500, which meant a shortfall of 8,000. Thus, to bridge this gap, a plan was launched to recruit science and engineering students. In October 1957, the USSR successfully launched the first satellite, Sputnik, into space. The shock of the Soviets' engineering prowess also raised worldwide awareness of the need to promote science and technology.

Despite the steady progress, the shortfall of scientists and engineers was estimated to reach up to 170,000[2] during the 1960–1970 period, as covered in the Council for Science and Technology's (CST) advisory report no. 1, *On the Comprehensive Basic Measures to Promote Science and Technology Targeting Ten Years Later*, and the national income doubling plan (*kokumin shotoku baizō keikaku*). Therefore, the plan was expanded to create capacity for an additional 20,000 students over three years from 1961. Measures were also adopted to address the rapid increase in university applicants by postwar baby boomers. Thus, the number of science and engineering students, which had been approximately 100,000 in 1960, exceeded 300,000 in 1970. In engineering faculties alone, the number increased from 70,000 to 220,000.

Since the shortfall of engineers became a social issue, an "industry-academia cooperation boom" and a "corporate school boom" (an establishment boom of the industry's in-house technical high school for employee candidates) were both established to complement the shortfall (Kitami 1962, 10; Hiroshige 1961). The industrial interest is evidenced in the activities of the Japan Science Foundation (JSF), established in 1960 mainly by Keidanren (the Business Federation).[3] This foundation established the Industry-Academia Cooperation Committee as "a research and deliberation organization to promote collaboration between the industry, the university, and public and private research institutes and support industry requests for research and re-training" (JSF 1989, 38). They aimed to (1) resolve the shortfall of engineers, (2) promote important research, (3) determine research themes, and (4) examine the research system of Japan. In particular, in 1961, they built the Industry-Academia Cooperation Center in Tokyo to resolve the shortfall in engineers. However, due to the challenge in realizing the goals, the center could not live up to its full potential, and it was only possible to continue holding workshops and seminars. They then donated the center's building to the Science and Technology High School (Kagaku Gijutsu Gakuen), established by the JSF in 1964. Their activities were, thereafter, continued at the Science Museum of the JSF (JSF 1989, 87–89).

Other business associations were also involved in industry-academia cooperation during this period. In December 1959, when Keidanren decided to establish the JSF, the Technology Education Committee of Nikkeiren requested the Liberal Democratic Party to establish specialized colleges for the training of middle-level engineers, which were equivalent to the higher technical schools of the prewar system (*Nikkei Shimbun* 1959a). Keizai Doyukai (the Japan Association of Corporate Executives) announced the establishment of the Industry-Academia Cooperation Center (as an alternative to the Industry-Academia Cooperation Committee of the JPC) to eliminate the overall shortfall of engineers and increase their quality (*Nikkei Shimbun* 1959b). Moreover, the Japan Iron and Steel Federation decided to contribute 250 million yen in funding over four years to the science and engineering faculties of national universities to produce more engineers.

In addition to the measures taken to resolve the shortfall of engineers, industry-academia cooperation provided financial support to universities to benefit companies with technological innovations. As the student recruitment plan began to succeed, technological innovation became a key cooperation factor (Kitami 1962, 9). Notably, Keio Engineering Foundation was established in 1959 as an affiliated organization of the Faculty of Engineering at Keio University to arrange commissioned research. In 1960, the Himeji Science and Technology Center was established to promote regional cooperation in the Harima industrial area. Furthermore, Toray Science Foundation was established to fund academic research.

Industry-academia cooperation began with the need to improve engineering education in postwar Japan. However, it was only ideal for some engineering professors and became a convenient slogan for addressing the shortfall in engineers and technological innovation based on business association interests. However, while the industry requested academic cooperation to remedy their urgent needs, they paid little attention to the American sandwich program of engineering education, which would require systematic cooperation as an indispensable ingredient. National policy also followed this trend. For example, the CST's advisory report no. 1 stated that "industry-academia cooperation research is important not only to create brand new technologies by connecting the basic research of universities and public research institutes with applied research and technical practices of the industry but also to cultivate and re-train researchers in business corporations." It also suggested that joint research facilities must be established, and commissioned research must be facilitated. Their vision aligned with that of the JSF without mentioning the sandwich program (CST 1960, 209). This industry trend was also consistent with the interests of private universities in forming and expanding their management bases. The Faculty of Engineering at Seikei University was established in 1962, and Kanazawa Institute of Technology (KIT) and Kyoto Sangyō (Industrial) University, 1965, all adopting industry-academia cooperation as their mission statements. They, however, did not introduce corresponding curricula of industrial cooperation.

The sandwich education system by industry-academia cooperation placed a heavy burden on private companies, and most companies had neither an interest in nor an understanding of the system. Following the Fourth JSEE Annual Conference in 1956, Wada Masao, the executive director, criticized the attendees' impression of "great success" of the conference as follows:

> I may be perverse, but I protested because I did not think it was a great success. I would be delighted with great success if half of the attendees, or a third at least, or a quarter even, were people from the industry. "What's this? It's a crowd of college teachers!" The JSEE is an organization for collaboration between industry and university, not an event where teachers gather money from the industry and meet together. [. . .] The brain of the industry is full of realistic problems; running a business and maintaining the economy of factories; in a word, making money. In the comfortable "white palace" universities, despite having little maintenance budget and looking like "dirty ruined temples" relative to company buildings, the professors receive their salaries from our nation, preach liberal arts and other academic studies to their students, and are dominated by issues on education. These two, after all, are different species from different worlds. (Wada 1956, 218–219)

This view of the companies was preserved throughout the 1960s, and the main topic of the cooperation shifted from engineering education to economic and industrial policy with more general and abstract concepts. For example, at the panel discussion of the Twelfth JSEE Annual Conference in 1964, Okoshi Makoto noted that, based on the result of a questionnaire survey of 509 companies (43.5% response rate), most companies undertook off-campus training "reluctantly by begging from the university or otherwise considering self-interests. In contrast, companies that accepted it from the broad perspective of industry-academia cooperation were much fewer than we expected." In addition, although he expected that the sandwich program could be implemented, at most, only for third- or fourth-year students in the Japanese system, he was disappointed by the companies' misunderstanding of the program. Many companies accepted only fourth-year students whom they would employ, as they perceived difficulties with third-years due to their inexperience (JSEE 1965, 38–41).

As described in chapter 2, after his move from the University of Tokyo to Toyo University, Okoshi attempted to introduce industry-academia cooperative education modeled on the American sandwich program. The system at Toyo University was successful because it commenced with a donation of 200 million yen from Hitachi, Ltd. However, this success was exceptional, and it did not proceed in an ideal way. As the national plan of student recruitment became successful, the focus on industry-academia cooperation shifted to commissioned research for technological innovation. Thus, *industry-academia cooperation* lost its original meaning and became a generic buzzword.

COOPERATION AS A SIGN OF MONOPOLY CAPITALISM

Following the signing of the 1952 Peace Treaty of San Francisco and the Japan-U.S. Security Treaty on the same day, Japan entered a new planned economy period under the postwar security arrangements. The Japan-U.S. Security Treaty was revised in 1960, and the U.S.-Japan Cooperative Science Program was initiated, based on Prime Minister Ikeda Hayato and President John F. Kennedy's talks in 1961. From the Marxist-Leninist approach in the progress of the U.S.-Japan relations, the university reforms on the industry-government initiative were understood as a revival of the prewar, imperialistic *monopoly capitalism*. Industry-academia cooperation was regarded as a combination of scientists and economic production in the interest of capitalists, and the 1957 Central Council for Education report was understood to be a comprehensive substitute for the voice of capitalists by the advisory board of the MOE minister (Onuma et al. 1975, 228–230; Shibata 1966). This understanding was strengthened when, in July 1960, in response to increased demonstrations around the National Diet Building (Kuroda 1999),[4] Keizai Doyukai implemented a policy to utilize industry-academia cooperation politically and socially to attract students to the capitalist side. It announced the promotion of "a new theorization of capitalism to raise sympathy from the intellectual middle class" by communicating more with students and scholars through activities with young scholar groups and the Industry-Academia Cooperation Center and addressing political and social issues (*Asahi Shimbun* 1960).[5] In response to the subsequent increase in college student riots, Keizai Doyukai proposed the *Higher Educational System for Higher Welfare Society* in 1969, seeking the "establishment of moral value in the new society" as a policy for industry-academia cooperation (Keizai Doyukai 1976, 327–328). A subcommittee of the Japan Scientists' Association (1970, 22) cited this 1969 statement as evidence that industry-academia cooperation had an aspect of "ideologue training and ideology dissemination to advocate the capitalist system."

In the early 1960s, the Basic Act on Science and Technology and the university management system became major social controversies. The enactment of the Basic Act was proposed in the CST's advisory report no. 1, and the draft was submitted to the House of Representatives in April 1962. Shimizu Kinji (1960, 102–103), who was in charge of the Basic Act as chief of the subcommittee for this report, explained that the Act aimed to solve a tax system problem to procure financial resources. Nevertheless, apart from Section 5 (on engineering), the SCJ opposed the legislation (Hiroshige 1965, 134) and adopted a recommendation to enact the alternative Basic Act on Scientific Research in April 1962 that maintained the freedom of scientific research, respect for the independence of scientific researchers, and publication of scientific research outcomes in principle. Prime Minister Ikeda's statement in May 1962 drew attention to the MOE minister's veto over university personnel affairs and the authority and autonomy of the administrative council and faculty council in university management. The minister requested the Central Council for Education for an advisory report on university management, *On the Improvement of University Education*, in 1960. Professors and students protested when they interpreted this "improvement" to the university management system as state intervention in the universities' autonomy and academic freedom.

Furthermore, in the 1960s, two successive events woke the public's social consciousness: expansion of the plan to outsource the R&D expenditures of the Technical Research and Development Institute at the Japan Defense Agency and the introduction of U.S. Army funds to the International Conference on Semiconductors held in Japan. Military-academia cooperation was criticized by those seeking academic democracy, autonomy, and openness; it was regarded as an expansion of national monopoly capitalism in the relationship with the defense industry (Kawasaki 1969).

Industry-academia cooperation, thus, became a target of criticism of ideological *monopoly capitalism*. Initially, it seemed like a vague criticism of the abstract relationship that did not reflect reality.[6] Taketani (1961, 42) criticized the attitude of the industry, saying that "entrepreneurs ignore it as *begging money* or an event somewhere else and treat it as if they were religious groups collecting donations." He concluded that "as far as science and technology are concerned, we can never find the *powerful will of monopoly capital* at all. We can only see hectic behavior so as not to miss the bus of technology tied up with advanced countries."

However, in the late 1960s, "the oppression of humanity, the phenomenon of human-alienation [*ningen sogai*], poverty, war, and the political corruption behind prosperity" in the highly industrialized society became more apparent and fundamental criticism of the usefulness of academic activity was undertaken, together with criticisms of giving the highest priority to economic growth and research funds acquisition. Such criticisms spread globally as the *student class* activism emerged alongside the popularization of going to university (Preparatory Investigation Committee 1969, 3–17). While student activists addressed global issues, such as opposition to the Vietnam War, they also targeted social contradictions and oppression within their countries. In Japan, criticism of industry-academia cooperation became an important issue. Student activists in Japan believed that under the Japan-U.S. security arrangement, Japan was complicit in U.S. imperialism, as manifested by the Vietnam War. Industry-academia cooperation was also understood to be part of *national monopoly capitalism* in the *imperialist reorganization*. This understanding resulted in a protest movement against the renewal of the Japan-U.S. Security Treaty in 1970.

A slogan, "objection to the policy of industry-academia cooperation," was first seen in the struggle at Waseda University at the end of 1965 to reverse the tuition increase and obtain management rights over a new student union building. Meanwhile, students campaigned against the university's all-around reform "to meet the demands of the national society." On the one hand, the administration attempted to expand engineering, management, and teacher training courses. On the other hand, they attempted to abolish "useless" evening courses and the newspaper and local government faculties. In this respect, the educational administration itself was regarded as representing industry-academia cooperation, which became an abstract slogan toward "universalization of individual struggles" (Record Editorial Committee for the Waseda Struggle 1966, 13–31).[7]

The institutionalization of industry-academia cooperation to redress the shortfall of engineers was criticized in the Waseda Struggle. Students perceived the administration reform's aim to change their university into a factory for mass-producing human resources in response to industry demands and promote non-elitism of students regarding employment. Suga analyzed this sentiment as follows:

> Industry-academia cooperation was opposed not because it promoted an *imperialization* of Japanese capitalism but because it was a capitalist propaganda that made it impossible for students, who could not help being produced as relative surplus personnel, to wipe out their unemployment fears. (Suga 1998b, 106)

The criticism of industry-academia cooperation in the mid-1960s emerged as students, who had a keen sense of their rights due to the postwar democratic education, claimed that their interests were being eroded by the interests of industry and university administration. Therefore, their criticism did not target a specific method of engineering education but the entire ethos of the administration.

COOPERATION AS A PROBLEM OF ACADEMIC INDEPENDENCE

In the University of Tokyo (Tōdai) Riots from 1968 to 1969, industry-academia cooperation was even more radically criticized as a symbol of the dilemma inherent in the requirement to secure the autonomy of and meet social demands for academic research and education at universities (Preparatory Investigation Committee 1969, 10–12). The lack of autonomy perceived by the students meant that "academic research may be subordinate to the interests of sponsors" and "the choice of the research plan will be influenced by factors incompatible with the original demands of the academic research" (Preparatory Investigation Committee 1969, 99–100). Thus, this criticism of university administration developed further into a criticism of the value of academic research.

Separation from social demands meant that the university had become exclusive and dogmatic; an *ivory tower*. Another problem was equating industrial demands with societal ones. After the war, it became obvious that "the character and function of universities have changed drastically, and the university as a social system has appeared more powerful than the so-called *ivory tower*" (Central Council for Education 1963, 2). Thus, it seemed apparent to the professors that it could not remain an "ivory tower" (Preparatory Investigation Committee 1969, 11). Accordingly, in 1969, the Council of the University of Tokyo published a policy: "We should deny *industry-academia cooperation* if research at the university loses its autonomy by serving the interest of capital" (Kato 1969, 9).

Hence, engineering professors organized the Faculty of Engineering's Investigation Committees and began discussing faculty reforms. Mukaibo Takashi (1969, 120–121), the dean of the faculty, referred to the issue of *denial of industry-academia cooperation* in his resignation speech: "In the Faculty of Engineering, along with the development of academic technology, a change from the *ivory tower* to intimate collaboration with society in the form of industry-academia cooperation has occurred naturally." He then went on to describe the direction of engineering as follows:

> Today, without cooperation with the industry in various ways, it seems impossible to advance engineering research and education. And it is quite natural that the outcomes of engineering research at the university will be beneficial to the industry. However, some industries that developed from engineering produce environmental pollution, and the development of the industry, in the words of students, will serve the monopoly capital and strengthen state power. In short, how will the results of the development of engineering affect society, and how will the benefits be distributed in society? Without reflecting on such problems, is it acceptable to conduct engineering research while staying in peace in the *ivory tower* or strengthen cooperation with the industry? Development of engineering will be prevented if we are trapped only in such a problem. However, the sound development of engineering and society in the future must take such criticisms honestly and reflect on them modestly, isn't it? (Mukaibo 1969, 121)

While Mukaibo affirmed industry-academia cooperation, he could not provide specific directions regarding social problems, which was the dilemma and basic directions of the faculty reform. At that time, four major lawsuits on environmental pollution and diseases were in the court: Niigata-Minamata disease (mercury poisoning) and Yokkaichi asthma (air pollution) in 1967, Itai-itai disease (cadmium poisoning) in 1968, and Minamata disease (mercury poisoning) in 1969. Such cases were becoming a problem in Japan. Other engineering professors also considered the dilemma; nevertheless, the solutions came from the narrow perspectives of engineering. They did not extend their discussions to the general problems of society, such as environmental pollution. They reassessed industry-academia cooperation based on their convenience; therefore, the dilemma remained.

The Faculty of Engineering's Investigation Committee expressed their opinions on industry-academia cooperation. Kondo Kazuo (1969, 135–136), an Investigation Committee H[8] member, charged with examining the "ideal of the university," noted that "in any sense, the university must never be a slave or parasite of the industry," and industry-academia cooperation was "merely a reference for both to strengthen their individuality." Watanabe Shigeru (1969, 152–155), also a committee member, however, argued that the policy of the Confirmation with the Delegates of Seven Faculties was that "the article itself is meaningless." He said that university research could not advance to a new academic and systematic knowledge without real issues of private companies since advanced research topics, even those that can only be treated at universities, were latent in the actual production process. In addition, he defended industry-academia cooperation because the number of laboratories in Japanese universities and that of companies seeking technical development were similar, and the connection provided graduate employments. Defending the university's position in a capitalist society, he stated that the argument that academics should not serve only one company was not necessarily valid.

The university-wide investigation committee pointed out that the financial support for research and education from the national government was too low and could be seen as a fundamental factor in academic nonindependence in the industry-academia cooperation (Preparatory Investigation Committee 1969, 99–100). At the University of Tokyo, the total budget per professor greatly increased from 1957 to 1960, from which there was little increase. There had been no increases in the research and management budgets of the Faculty of Engineering since 1967. Therefore, the gap between the growth of the gross national product and research investments widened (Inagaki et al. 1971, 150–151).[9]

The lack of financing caused a dependence upon commissioned research, which created a significant strain, particularly for graduate students. For example, graduate students at the Department of Urban Engineering at the University of Tokyo published their activities in the Japanese magazine *Shizen* (*Nature*) in April 1969 (Graduate Students Union of Urban Engineering 1969). The Committee of Education of the House of Representatives criticized the fact that the contents of their research and education were, in reality, subcontracted works of commissioned research (Committee of Education 11 (61st Diet Rec. Apr. 11, 1969)). In particular, the students considered it a great dilemma that their department's commissioned research, such as the Narita New Town planning and the Tsukuba Science City planning,[10] was the cause of major social dispute. Aizawa Yoshikane, a graduate student at the Department of Urban Engineering, explained this situation:

> Most research in urban engineering is at a very early stage of academic level, so we have to first engage in practical activities. Since it was a stage to generalize the research gradually over time, it remained only at the level of easy practice. Thus, commissioned research was introduced. Most of our curriculum comprised subordinating ourselves to commissioned research because most of the

research depended on it for funding or other reasons. This situation means that when entering the graduate school and laboratory, one is obliged to do commissioned research. (Aizawa et al. 1969, 85)

The root of the industry-academia cooperative problem lay in the funding issues related to the expansion and popularization of universities. Moreover, there was a conflict between ideal and traditional academic values and the real problems universities faced. Suga sees the fundamental contradiction in the criticism of industry-academia cooperation at the time:

I have doubted since then: even if the viewpoint of "objection to industry-academia cooperation" was certainly effective in revealing the political nature of academic activities, at the same time, it was a double-edged sword that further strengthened the myth of value-neutrality of academic activities. It would be the only possible way to shut themselves up more deeply in their *ivory tower* or to do nothing, so that their own academic activities would not be utilized by capitalism (and politics as well). At least, that is relatively more effective. This situation is *self-denial*, which is the only possible way for those who are in capitalism. It is still my unaltered recognition that the consequences of the catchphrase "self-denial," which was shouted by the All Students Joint Struggle Conference at the University of Tokyo [Tōdai Zenkyōtō] and was touted in journalism at the time, would become only such a thing after all. In fact, neither "objection to industry-academia cooperation" nor "self-denial" was realistic, in the sense that they were merely obsessive and repetitive representations of a classical myth of the university's academic values—academic activities *for the people*! (Suga 1998a, 105)

The justification of industry-academia cooperation was encouraged by incorporating the historical analyses of a "classical myth." Aoki Seizo (1969, 17–22), a historian of science, explained that the universities in medieval Europe (law, divinity, and medicine) were products of a similar idea of industry-academia cooperation created by social demands, and the establishment, preservation, and maintenance of their legitimacy were required by becoming academic. Modern science, according to him, was also developed by mechanical technology outside the university, which then became conservative by being incorporated into the university. Quoting Aoki's analysis, Kihara Junji, at the Faculty of Engineering at the University of Tokyo, urged for a reflection on the "vacant authoritarian attitude" because "we must check ourselves again for whether our academic activities have become self-purposed and protect ourselves from such degradation" (Kihara 1969, 360). Such identification of industry-academia cooperation with social demands universalized the interests of the industry, particularly the interests of specific industries or companies with political power, to the interests of the society. This situation could not resolve criticism that industry-academia cooperation was subordinated to individual interests. However, on the promoting side, this understanding shaped the recognition that industry-academia cooperation should be the ideal form of academic activity and further urged the promoters to deny the critical attitude and justify themselves.

Ui Jun (1985, 25), who worked heretically on the social issues of environmental pollution at the Department of Urban Engineering, recalled: "Of course, most students and graduate students at the Department of Urban Engineering disappeared, the professors regained their confidence, and a more conservative atmosphere dominated the department than ever before." Against such an atmosphere, he began two series of voluntary open lectures titled "Kōgai genron" (Principles of Industrial Pollution) in 1970 and "Daigaku kaitairon" (Dissolving University) in 1974. Some causes can be assumed in the "conservative atmosphere" noted by Ui. Particularly, the professors re-examined themselves and deepened their confidence in the conventional attitude of the faculty, which had promoted industry-academia cooperation. For example, Yoshikawa Hiroyuki (1990, 15–19), the youngest associate professor at the faculty, recalled that students accused him of production (industrial) engineering in a "confined" discussions with them during the riots: "Why does the university do something like production to lead to corporate interests?" He then figured out his research theme of the general design theory by referring to the teachings of Okoshi Makoto and artificial intelligence as the intellectual activity of the human being behind engineering design.

In 1961, Hiroshige argued for the necessity of revising the conventional, conservative view of academic activities. As the focus of science and technology research had shifted to the private sector, companies had less need to commission their basic research from universities, and industry-academia cooperation was required in business management rather than industrial technology. Furthermore, considering that the mutual isolation of science, technology, and industry was the "most remarkable defect" of science and technology in Japan, he claimed:

> If a university, without initiative and independent vision for overcoming the defect, could bring up nothing as a counter concept but classical concepts such as the autonomy of the university, universities would be swept away quite passively in the tide of industry-academia cooperation. (Hiroshige 1961, 110)

His cautionary observation became a reality in the 1960s. However, it seems that the objection was never effectively overcome. Moreover, due to the depth of the dilemma that led to its "self-denial," industry-academia cooperation came to be regarded as a "taboo."

REACTIONS FROM THE PROMOTING SIDE

In contrast to concerns about academic independence, the promoting side advocated the science, technology, and industry interrelationship as a proactive way to overcome the problems. In February 1969, Keizai Doyukai announced its "basic views on university problems we face" and once more defended industry-academia cooperation. Moreover, in the proposal titled *Higher Educational System for Higher Welfare Society*, published in July of the same year, they insisted cooperation was a trend in world history not only because it was "consistent with modern educational thought that shifted from *academic activity for academia* enclosed in an ivory tower to *academic research opened to society*" but also because countries such as the United States, Russia, and Japan achieved industrial advances in a short period. They also expected that "the progress of R&D in companies, the advancement of research institutes outside universities, the emergence of big science, and the rise of social demands for continuing education will increase the importance of industry-academia cooperation in the future" (Keizai Doyukai 1969, 18). In this proposal, they advocated that "dedication and service to a variety of belonging societies from family to humankind" were important because "immature individualism prone to self-righteousness and egoism cannot be a pillar of the ethics of higher welfare societies" (Keizai Doyukai 1969, 11). This mode of ethics is similar to that of their 1960 announcement, which was criticized for drawing students toward capitalism.

In 1972, the Tokyo Chamber of Commerce and Industry (1973) conducted a questionnaire survey of several hundred people, including business managers and university presidents, on industry-academia cooperation. They stated that the term *industry-academia cooperation* was somewhat a taboo or allergen after the college riots, but the institutionalization of massive cooperation must be attempted in response to the new era. From the results, 92 percent of respondents answered that the promotion of industry-academia cooperation was "necessary," including 85.9 percent of university presidents. Only 1.4 percent answered otherwise. The survey also investigated the introduction of the "sandwich" education program, modeled on the United States. Accordingly, 70.1 percent of the total and 79.1 percent of university presidents answered that it was "useful." Furthermore, 57.5 percent of the total answered that such an education was necessary for the commercial field.

In response to these results, the Tokyo Chamber of Commerce noted that a large part of the industry-academia cooperation had always been restricted to

R&D within a few large companies; the cooperation had not been institutionalized and the present situation was that "the majority of small- and medium-sized companies cannot grasp even a clue of cooperation." Furthermore, "there are criticisms of the unclear transfer of funds in research cooperation between the university and companies, of the loss of independence of research and a neglect of education at university." They concluded that "these criticisms caused allergies to industry-academia cooperation and have become a factor that impedes the spread and promotion of such an institutionalization of cooperation."

Evidently, the Tokyo Chamber of Commerce and Industry did not consider the sandwich education system to be widely known because the related survey question was annotated with an explanation of the system. Even in the 1970s, the institutionalization of industry-academia cooperation did not progress much, as the concept changed significantly during the 1950s and 1960s, making it more ambiguous and challenging even for the promoting side to understand. Industry-academia cooperation was mainly promoted by specific industries and large companies with economic and political power. Meanwhile, for most small- and medium-sized companies, it was challenging to undertake. It was also important for the promoting side to reduce the bias in each case. As the promoting side pushed for industry-academia cooperation abstractly without resolving the criticism of it, the opposition side repeatedly insisted that academia should be independent.

In this situation, the JSEE also cautiously promoted industry-academia cooperation. In 1969, Wada Masao (1969, 37), vice president of the JSEE, described a feature of industry-academia cooperation in Japan: "Both the industry and the university are not in a cooperative system; the university leaves it almost entirely up to the industry, and the industry is not willing to accept trainees but accepts them reluctantly by thinking of it as an unavoidable obligation demanded by the university." He then criticized such attitudes of companies as follows:

> This is a really strange story. The industry must understand that the university is taking care of engineers who will be their life-long employees as a student for four years to provide basic education. If the industry's demands to students cannot be fulfilled by the university, but they can be fulfilled by the industry, it is natural that the industry should educate students. This initiative will benefit the industry itself. (Wada 1969, 37)

Although Wada expressed dissatisfaction with companies' disinterest, lack of understanding, and noncooperative attitudes, other engineering professors failed to propose any concrete improvement plan, despite agreeing with him in principle. At the second meeting of the JSEE Committee of Educational

Policy in December 1968, Wada's draft of the Declaration of the JSEE's Intention was circulated and the contents were discussed at the third meeting in January of the following year. Wada's draft stated the following: "4. Since this society is a consultation of industry and engineering professors, we declare that the industry-academia cooperative system is the constitution of the society" and "5. Since this society aims for the development of human resources as a foundation of industrial development, we declare that we are daring to establish *industry-academia cooperative education*." In the fourth meeting in February, the opinion, "must be persuasive because it is quite difficult to convince young people that industry-academia cooperation is necessary," was expressed on this proposal and "do not deny industry-academia cooperation" was settled on (JSEE 1969, 90–94). The following statement was expressed at the fifth meeting in March:

> Although there is no objection to the assertion that society must strongly declare industry-academia cooperation, this issue must be treated with caution because insisting on industry-academia cooperation is likely to cause trouble to the member professors at present. No one opposes industry-academia cooperation in the industry. Nikkeiren has clearly announced its support to industry-academia cooperation. They can say it clearly because there are no university members. For our society, it is necessary to clarify the aim for industry-academia cooperation, but many members agree with the opinion that they want to wait for a while to consider the means of declaring this support; however, we should not become allergic to industry-academia cooperation. (JSEE 1969, 94)

The JSEE felt there was little resistance to industry-academia cooperation from the educational side. Primarily, criticism in research was narrowed down to how research funds were managed, particularly by focusing on the problem of professors who accepted commissioned research as a *backdoor business* without official university management and urged students to conduct it. In the interim report of the committee, the JSEE policy was summarized as follows:

> It is necessary for both the industry and the university to recognize the social mission in each position and cooperate closely. Needless to say, each request for cooperation should be responded to at their own discretion. It was only questioned as an abuse of industry-academia cooperation due to the insufficiency of attention to clarify the usage of funds in the traditional research practice of the cooperation. For industrial development, industry-academia cooperation is increasingly required, where both sides consider and implement projects voluntarily based on each social mission by establishing *rules* to clarify the usage of funds. (Kodama 1969, 96)

The panel discussion titled "The Necessity and Concrete Measures of Industry-Academia Cooperation in Engineering Colleges" at the JSEE Annual Conference in 1970 took a similar direction to the committee's discussion. After speeches by each panelist, an attendee noted the lack of mutual understanding between the industry and the university, while the moderator supported the criticism by saying, "99 percent is exaggerated, but I think that quite a lot of industry and academic people are disinterested," among various opinions expressed (JSEE 1970, 36). Eventually, all panelists expressed themselves. However, they failed to formulate any concrete measures during the three-hour discussion. Therefore, an abstract and vague conclusion was obtained, such as a lack of communication between the industry and the university.

Consequently, engineering education reform via industry-academia cooperation, which was expected as an ideal exemplar from the United States in the advent of postwar democratization, had not been the powerful vision it was once thought to be. Industry-academia cooperation was considered a taboo at universities, and the conventional setups of university faculties were preserved. There was little room for seeking vocational education for professional engineers at the university, because it was considered the role of technical schools. In the 1980s, when Japan's university system became a social issue again due to the anachronism amid the serious trade friction between Japan and the United States, changes were sought in the industry-academia relationship and university education. As the changes accelerated with globalization in the 1990s, engineering ethics came to be required as an important factor. The following chapters describe the details.

NOTES

1. The Faculty of Pharmaceutical Sciences also did not sign it, and the Faculty of Economics later withdrew their signature.

2. This number is significant relative to the 8,000 in 1957 because the 1960 report was for various scientists and engineers, while the 1957 report was for engineers only.

3. The first chairperson was Kurata Chikara, president of Hitachi, Ltd., who donated to establish the Faculty of Engineering at Toyo University.

4. On June 15, 1960, the demonstration escalated, protestors burst into the building, and an accident occurred in which Kanba Michiko, a fourth-year female student at the University of Tokyo, died.

5. This announcement was made by Kikawada Kazutaka, chair of the association. A round-table discussion of the Engineering College Students' Union, which represented ten universities, including Nihon University and Tokai University, supported by the JSF and the Daily Industry News, Ltd. (Nikkan Kogyo Shimbun), expressed positive expectations of the foundation, and there was essentially no criticism of

industry-academia cooperation. They raised the issue that the science and technology in the industry was substantially more advanced than what was being taught in universities, and students were concerned about the gap when starting their employment. As a countermeasure, they proposed not only sharing information between industry and students by establishing an intermediate foundation but also expanding educational facilities, including vocational education, and disseminating scientific and technical knowledge through co-activities (Engineering College Students' Union 1960).

6. According to the survey conducted by the JPC, companies that did not know the commissioned researcher system comprised 24 percent large companies and 55.5 percent small- and medium-sized companies. Furthermore, 79.8 percent of all the companies did not use it despite knowing about it (97 percent of small and medium companies). Additionally, two-thirds of large companies conducted commissioned research for universities, but only 17.4 percent of small and medium companies did so. Therefore, the JPC concluded that there was a need for further PR because the awareness of industry-academia cooperation remained low (JPC 1960).

7. The specific understanding of the industry-academia cooperation differed among the student parties (Record Editorial Committee for the Waseda Struggle 1966, 19, 293).

8. When the new administration of Dean Mukaibo was established in November 1968, Investigation Committees A to H were organized for each theme to examine and report various problems in the Faculty of Engineering to the faculty council.

9. The Subcommittee of Japan Scientists' Association in the Faculty of Engineering at Kyoto University (1970, 19–21) also criticized the unprincipled expansion that blindly followed industrial dictate (by opening a course of almost identical contents and changing the name) and distortion in the contents of research and education that violated the principles of democracy, autonomy, and openness by introducing corporate funds. They also noted that there was a national "poverty making policy" for university budgets.

10. These plans were for the construction of Narita International Airport, the University of Tsukuba, and the national research institutes.

Chapter 5

The Growth of Industrial and Practical Demands

RELUCTANCE TO ENACT A CODE OF ETHICS FOR RESPECTIVE ACADEMIC SOCIETIES

A worldwide scientists' movement emerged after World War II to reflect the significant increase in the social influence of science in the twentieth century, thereby promoting the establishment of codes of ethics. The World Federation of Scientific Workers (WFSW), a labor and peace movement organized in 1947 to improve the social status of scientists, established the Charter for Scientific Workers in 1948. Furthermore, the International Council of Scientific Unions established the Fundamental Charter of Science and the Charter for Scientists in 1949. Moreover, with the growing threat of global nuclear war evoked by the hydrogen bomb test at Bikini Atoll in 1954, the 1955 Russell–Einstein Manifesto and the following Pugwash Conferences on Science and World Affairs strongly demanded scientists' social responsibility in deterring the development of nuclear weapons that could destroy humankind. Furthermore, given the increasing environmental pollution, big science (high-energy physics and radio astronomy), and life sciences from the 1960s to 1970s, ethics in science became necessary to prevent abuse of scientific knowledge and control the outcomes of scientific activities (SCJ 1972, 37; Oki 1978, 4–5). Hence, the United Nations Educational, Scientific and Cultural Organization (UNESCO) adopted the Recommendation on the Status of Scientific Researchers in 1974, a Bill of Rights for scientists that defines their social responsibility as a profession (Okakura 1980, 27–29).

In postwar Japan, the Association of Democratic Scientists, a large communist group of Japanese scientists, joined the WFSW and developed scientific enlightenment initiatives. Twenty years later, the Science Council of Japan (SCJ) requested that the government implement the UNESCO

Recommendation in Japan in 1975 and made the second recommendation for enacting the Basic Act on Scientific Research in 1976. The SCJ also requested the realization of the Basic Act as a democratic law for scientists in 1962 in opposition to the Basic Act on Science and Technology proposed in a national policy. The SCJ's Basic Act was a claim of the state's responsibility to scientists; that is, the scientists' rights to the state. Therefore, in 1980, The SCJ established the Charter for Scientific Researchers as a professional code for scientists to declare their social responsibility and obligations and be introduced alongside the rights (SCJ 1980; Fushimi 1980, 17–18; Igasaki 1980, 73).[1] The SCJ Charter called for scientists to keep their minds pure and disciplined in favor of truth, contribute to human welfare and world peace, respect the freedom and harmonious development of scientific research, and eliminate the disregard and abuse of scientific research.

In Japan, although professional ethics for *engineers* did not become common despite the efforts of some leaders, the social responsibility of *scientists* was promoted per global trends. However, the code was written for the community of scientists, not for respective academic societies. Individual academic societies established their codes of ethics in the United States. However, the culture was different in Japan. For example, in addition to the 1974 UNESCO Recommendation, the 1966 ILO and UNESCO Recommendation Concerning the Status of Teachers was often cited as a reason for establishing the SCJ Charter. In contrast, the code of ethics for physicians, adopted by the Japan Medical Association (JMA) when joining the World Medical Association (WMA) in 1951,[2] was never mentioned in the SCJ's discussion. In medical ethics, only the Hippocratic Oath was mentioned. This difference stemmed from the strong influence of UNESCO (an international organization common to all disciplines) and the fact that the JMA was a professional organization in a specific field.

The reluctance to establish a code of ethics for individual societies appeared in the case of the Institute of Electrical Engineers of Japan (IEEJ). Kimura Hisao, the chairperson of the IEEE Computer Society Japan Chapter and professor at Seikei University, introduced the IEEE's code of ethics in the journal of the IEEJ in 1976 and received favorable comments. Thus, he proposed that the IEEJ should establish a code modeled after the IEEE. In 1973, Kimura was shocked that a report on the ideal image of engineers, deliberated on for over a decade by an academic society's committee, had only concluded that "we do not know what to say after all" (Kimura 1976, 769). Meanwhile, in the United States, the IEEE revised the code of ethics to emphasize engineers' responsibility to society, following the internal whistleblowing on the San Francisco Bay Area Rapid Transit (BART) in 1972. Kimura was strongly influenced by the fact that each engineering society had its code of ethics.

However, in response to Kimura's proposal, an IEEJ board member demonstrated an "adverse reaction," stating that it was unnecessary to establish basic principles via a code of ethics for each academic society. Fearing that such an act would induce engineers to demand their rights in return for their social responsibilities, he stated that when a code of ethics is considered, it should be considered among all relevant societies, not in a single society. The board member also stated that he could not receive any response to his inquiry to the "academic society" (Kimura 1979, 44–45).[3]

The IEEJ's response to Kimura's proposal was probably conceived under the preparation of the SCJ Charter, which covered all scientific researchers, including engineering professors, demanding the rights of the entire scientific community. The IEEJ board member's "adverse reaction" was because the establishment of the code of ethics by each society could be regarded as a selfish act that disturbed the unity of the scientific community.[4]

Referring to the *Legal and Ethical Phases of Engineering* by Francis C. Harding and Donald T. Canfield (1936), Kimura maintained that the postwar Basic Act on Education was only a *legal phase*; no *ethical phase* like the prewar Imperial Rescript on Education was provided. He assumed that both the lack of understanding of *popular sovereignty* and the academic community's reluctance to enact a code of ethics were caused by the GHQ/SCAP's abolishment of the Imperial Rescript on Education as a prewar ultra-nationalistic moral code, which enhanced knowledge-based education in postwar Japan (Kimura 1977a, 37–40; 1977b, 250; 1978; 1979). Kimura's speculation was based on Japan's general complaint against the GHQ/SCAP-forced postwar education reform. In reality, the postwar reform introduced democratic education with popular sovereignty. However, as democracy became standard for the Japanese people, an opposite understanding was created in the 1970s.

Nevertheless, Kimura's claim saw no active support or objection. He was disappointed by the poor response, stating: "It seems that engineers in our country are not aware of the direct relationship with democracy and have distanced themselves from considering it" (Kimura 1979, 43). Afterward, Kimura continued to promote awareness of this ethical issue. He launched and chaired the Tokyo Chapter of the SSIT (SIT Tokyo) in 1983, following the IEEE's establishment of the Society on Social Implications of Technology (SSIT) in 1982 and held seminars on engineering ethics with the JCEA's Computerization Study Group (Suguri 1999, 58).

The reluctant attitude toward a code of ethics also persisted in the Japan Society of Civil Engineers (JSCE, formerly the JCES). Sato Mitsuharu, an engineer at Hitachi, Ltd. and a Gijutsushi, introduced the ASCE's efforts in engineering ethics in the journal of the JSCE in 1979. He offered the criticism that Japanese engineers, less autonomous in ethics, were surprisingly ignorant about other countries' basic attitude to preserve ethics and that the JSCE's code of ethics was no longer enforced. He assumed the cause as follows:

> Speaking of ethics and moral education in the general postwar atmosphere in Japan, the term moral education reminds us of unpleasant experiences of prewar Confucian and militaristic education and might cause an allergic reaction to the happiness of enjoying freedom brought by the defeat in the war. Otherwise, even if a code of ethics or conduct were created, a feeling of resignation and distrust would be evoked by the understanding that it is merely like a *Buddhist scripture* and that the *principles* (tatemae) and the *real intentions* (honne) are different. (Sato 1979, 7)

Sato criticized the morality of the civil engineering business, stating: "There are duty and obligation under intimate relationships, but ethics and morals are considered irrelevant and inappropriate for the business" (Sato 1979, 7). He was also concerned that the severe restrictions on Gijutsushi's qualifications became mere requirements for the government's authorization and somewhat restricted Gijutsushi's autonomy (Sato 1979, 8–9). As Sato described, both the JSCE and the JCEA established the codes of ethics, but neither code became popular among the members due to the postwar atmosphere surrounding moral education.

The reluctance to enact a code of ethics does not immediately imply the moral inferiority of Japanese engineers. On the contrary, the improvement of Japan's international status in engineering circa 1980 led many Japanese engineers to take pride in their moral discipline. The appreciation of Japan's uniqueness was similar to the growing interest in Bushido after wins in the Sino-Japanese and Russo-Japanese Wars circa 1900. Imai Kaneichiro (1980), president of the Japan Society of Mechanical Engineers (JSME) and board director of IHI Corporation, argued that the 1980s should be an "age of cooperation" for engineers rather than competition. Thus, it was important to maintain awareness of their professional responsibility to make high-quality products. Having introduced the ASME's code of ethics, he maintained that Japanese engineers' high ethical standard in product quality was often attributed to the emphasis on loyalty to their company rather than their salary. Although this value judgment might have been challenging for Americans to agree on, the spirit of making good products must have been common to both countries.

Leaving aside the superiority or inferiority of Japanese morality, neither a code of ethics nor professional ethics education for engineers was introduced in Japan until the 1990s. Engineering ethics was only sporadically noticed by engineers affected by the U.S. situation, as in the postwar period. Their systematic implementation required external pressure to comply with the global standard, which became a serious problem in the 1990s. In the 1980s, however, Japan promoted a science and technology policy to transition from a "catch-up" to a "front-runner" country by expanding R&D competition worldwide and sought to improve engineering education per industrial policy. The development of industrial and educational policies was a precursor to the introduction of engineering ethics. The rest of this chapter discusses these processes.

PROMOTION OF BASIC RESEARCH

The 1980 SCJ Charter called for freedom of scientific research; however, scientific research was incorporated into the science and technology policy promoted by industry-academia-government collaborations throughout the 1980s, emphasized in the industrial R&D framework.[5] The restructuring of the national research system based on the industrial policy was conducted with the active involvement of engineering professors at the University of Tokyo, without consent from most of the other researchers (Uemura 1989, 179–188).

In Japan during the 1970s, economic growth slowed after the 1971 Nixon shock (the collapse of the Bretton Woods system) and the oil crises; the expenditures for social security and public projects increased; and the accumulated deficit worsened. Fiscal rehabilitation became a serious political issue. As a policy to address the economic predicament, the Industrial Structure Council (Sangyō Kōzo Shingikai), chaired by Doko Toshio, who had a career as the president of IHI and Toshiba and the chairperson of Keidanren, called for the "technology-oriented nation" (*gijutsu rikkoku*) as a slogan in its 1980 report on the trade and industrial policy. The Second Provisional Committee on Administrative Reform (Dai Niji Rinji Gyōsei Chōsakai, Daini Rincho), organized in 1981 with Doko as its chairperson, proposed comprehensive rationalization plans such as the separation of functions between the central and local governments and the privatization of the three public corporations,[6] following the basic policy of fiscal restructuring without tax increases via the power of the private sector. The government failed to introduce the consumption tax in 1979 and provoked the business community's disapproval by raising the corporate tax rate in 1981. However, it steadily advanced to zero-based budgeting for 1982 and minus-based budgeting in 1983 and beyond.

Industry-academia-government collaboration (*sangakukan renkei*), different from the previous industry-academia cooperation (*sangaku-kyōdō*) since the former was a national policy, was introduced in the reports (RIITI 1980, 96–97; Uemura 1989, 109–113). The Second Provisional Committee's 1982 report formulated the strengthening of industry-academia-government collaboration as a fundamental approach to science and technology policy:

> In science and technology research in Japan, comprehensive and efficient promotion of R&D is insufficient due to the deficiency of voluntary and creative research and the sectionalism of the promotion of ministries and agencies. In order to overcome these problems in the future, it is necessary to establish a consistent promotion system with clear segregation of duties between public and private sectors, universities, and national institutes in science and technology research. In addition, more efficient and focused R&D should be promoted by facilitating industry-academia-government collaboration and utilizing private sectors' R&D capabilities. (Daini Rincho 1982, 37)

National universities were inferior to private companies in research equipment because, despite aging laboratories with crowding spaces and increasing research equipment expenditures due to research advancements, the governmental budget for research at universities had not been increased. Moreover, through the college riots in the 1960s, society doubted universities' management abilities. Universities kept a distance from the general public and were content with the isolated situation. Thus, to improve this situation, revitalizing universities to cooperate with society, in particular the industry, became a goal for policymakers. Furthermore, the promotion of basic research for industrial R&D at universities became a common interest in industry-academia-government collaboration. The first step was the procurement of private funds.

In Japan, the 1965 Survey on R&D introduced these research categories: *basic research* comprising *pure basic research* and *oriented basic research*,[7] *applied research*, and *development*. Meanwhile, the importance of systematic and international scientific research promotion, especially basic research, came under the spotlight through the UNESCO report in 1961: *Current Trend in Scientific Research* (*Tendances actuelles de la recherche scientifique*). This report categorized fundamental research (*recherche fondamentale*) into free fundamental research (*recherche fondamentale libre*) or pure research (*recherche pure*) and oriented fundamental research (*recherche fondamentale orientée*), further distinguishing it from application research (*recherche appliquée*), technical focus operations (*opérations de mise au point technique*), and industrial operations (*opérations industrielles*) (Auger 1961, 16–18). The OECD also held the First Ministerial Meeting on Science for science, technology, and innovation policy in 1963 and promoted fundamental research according to UNESCO's categorization.[8] Japan joined the OECD in 1964.

Japan's turning point for promoting basic research was establishing the Ministerial Council on Science and Technology in 1980, where comprehensive science and technology policy was emphasized. Based on this policy, the Special Coordination Funds for Promoting Science and Technology (Chōsei-hi) was established in 1981 to strengthen the comprehensive coordination function of the CST, and the following grant policy was introduced: (1) promoting advanced and basic technology; (2) promoting R&D with the cooperation of multiple institutions; (3) strengthening close industry-academia-government collaboration; (4) promoting international joint research; (5) flexible response to urgent research needs; and (6) conducting research assessments and surveys on R&D. Until then, the STA had promoted large individual projects in specific fields, such as nuclear power, space exploration, and ocean exploration. Moreover, in 1981, the Research Development Corporation of Japan launched the Exploratory Research for Advanced Technology (ERATO), a large research funding program to promote creative basic research, and the MITI introduced the R&D Program on Basic Technologies for Future Industries. Furthermore, the Grants-in-Aid for Scientific Research (KAKENHI), a general research funding program for universities, was increased for continuity (Kuniya 2015). Per the above categorization, the basic research required by these programs, except KAKENHI, was oriented toward basic research with the dream of industrialization.

Although basic research was promoted in the science and technology policy, the existing system of national universities and research institutions was not adapted to industry-academia-government collaboration. Therefore, the Research Cooperation Office was established in the Bureau of Science and International Affairs of the MOE in 1982; joint research between universities and private companies became possible via the MOE notice in 1983.[9] In 1984, the Science Council of the MOE (SCM, Gakujutsu Shingikai) also requested to establish research facilities to enhance the mobility of researchers, strengthen social cooperation and collaboration, and promote international exchange and cooperation in academic research (SCM 1984).[10] Moreover, the Second Provisional Committee announced a policy to abolish the nationality restriction for professors at national and public universities to promote the recruitment of foreign professors, which became possible via the Research Exchange Promotion Act in 1986. Thus, industry-academia collaboration has been consistently promoted as science, technology, and industrial policies, breaking the taboo described in chapter 4.

The pressure from the United States also strongly stimulated awareness of this basic research issue in the 1980s. The U.S.-Japan Cooperative Science Program, launched in the 1960s, was an academic exchange program that included all basic natural sciences and emphasized the autonomy of researchers while keeping the national interests of the economy and the military in mind. The U.S.-Japan Committee on Scientific Cooperation was launched for this program. Kaneshige Kankuro and Harry C. Kelly, who worked on the reconstruction of postwar Japan's science and technology system, served as the first chairs. Moreover, the Japan Society for the Promotion of Science (JSPS) and the NSF became coordinating institutions in each country and managed programs, such as seminars, joint research, and researcher exchanges.

At the fourteenth meeting of the U.S.-Japan Committee in 1977, the United States, which had begun to lose its industrial competitiveness with Japan, proposed to hold a symposium on the issues of both countries emerging from the progress of science and technology (such as the promotion of R&D and innovation). The proposal from the United States aimed for a mutual understanding of their science policy situations, not only in basic science but also in R&D. Thus, another opportunity was arranged at a governmental level (Onozawa 1981, 179–180). Accordingly, from September 29 to October 2, 1980, the First Seminar on Science Policy under the U.S.-Japan Cooperative Science Program was held at the East-West Center in Hawaii. In this first session, Harvey Averch at the NSF explained that embarking on science and technology policy was to activate its economic functions since it became clear that advances in scientific knowledge would create improvement in productivity and economic growth (Averch 1981, 182). Furthermore, no comprehensive coordination of R&D for the effective promotion of industrial policy existed in Japan, as this was becoming an important issue for Japan (Tezuka 1981, 187).

The United States directed the issues of the Seminars on Science Policy toward Japan's insufficient international contribution to basic research. Meanwhile, in the United States, the patent policy was promoted under the 1980 Bayh–Dole Act, and the enhancement of industrial competitiveness became a significant political issue, as noted in the Young Report, *Global Competition: The New Reality*, in 1985. At the OECD Committee for Scientific and Technological Policy (CSTP) Tokyo Meeting in 1983, Frank Press, president of the National Academy of Sciences (NAS), and Robert M. White, president of the National Academy of Engineering (NAE), called for Kobayashi Koji, president of NEC Corporation, and Inose Hiroshi, an electrical engineering professor at the University of Tokyo, both foreign associates, to discuss the issue of U.S.-Japan relations on advanced technology at an academy level. However, there was no such academy in Japan. Okamura Sogo, president of JSPS, responded by establishing the JSPS Committee on Advanced Technology and the International Environment (Committee 149). Committee 149, chaired by Mukaibo Takashi, comprised both leaders in industry and academia. Moreover, twenty-four major companies eventually became company members. Meanwhile, the U.S. Committee on Japan was chaired by Harold Brown of the Johns Hopkins Foreign Policy Institute (JSPS Committee 149 2000).

In August 1985, the First U.S.-Japan Meeting was held in Santa Barbara, California. The participants were divided into three working groups (research systems and the innovation cycle, telecommunications and data processing, and biotechnology). They discussed three themes (the differences between the American and Japanese systems, impediments to international trade and cooperation, and the national attitudes regarding basic research). At this meeting, the United States criticized the Japanese advantage over the United States in access to research results, which generally flowed from the United States to Japan, arguing that the pool of basic research should be a global, free commodity, and contributions and access should be universal and reciprocal. However, underlying this criticism were significant differences in the basic sciences and research infrastructure between the two countries.[11] While appreciating Japan's efforts to promote basic research in recent years, the United States argued that Japan should have more impact in world basic science research by promoting infrastructure and intercommunication (Inose 1986a, 6–9).

At the second meeting in November 1986 on the theme of the innovation process, the United States introduced the key phrase "symmetrical access." Okamura recalled this introduction as follows:

> What particularly impressed me at this meeting was that they proposed the concept of *symmetrical access* in response to Japan's counterargument at the first meeting: "Although the United States blames imbalances in information and human exchanges, isn't it because of the lack of your efforts, for example, to study Japanese, read Japanese papers, or put up with Japanese life to work at Japanese universities or research institutes?" (Okamura 1999, 290)

The United States first requested *equal access*. However, the resources required for innovation (knowledge, technology, capital, and markets) were not the same between Japan and the United States. For example, basic research in electronics was mainly conducted publicly at universities in the United States but was conducted privately in companies and national research institutes in Japan. Therefore, responding to Japan's opposition based on the difference between the two countries, the United States proposed *symmetrical access* as an alternative concept, carefully reworking the term *equal access* to demonstrate fairness to both countries (Inose 1987a, 86). After all, Japan and the United States decided to collaborate for several years to turn the asymmetry into a "positive-sum game" via a free market economy rather than protectionism.[12]

The criticism of asymmetry from the United States spread rapidly in Japanese society, along with the phrase "free ride in basic research" (*kiso kenkyū tadanori*). The criticism that Japan enjoys a "free ride" was common; thus, Japan was sensitive to it. For example, Johnson (1982, 15–17) described Japan as often being criticized as a beneficiary of the following three "free rides" in the postwar alliance with the United States: lack of defense expenditures, ready access to major export markets, and relatively cheap transfers of technology. Johnson discussed the significance of MITI's industrial policy after arguing that the first two were incorrect and the third was trivial and misleading. Moreover, Japan was often stigmatized as a technological imitator and free-rider (Okimoto 1986, 544). Harold Brown, the chairperson of the U.S. Committee on Japan, also accused Japan of taking a "free ride" when he was the Secretary of Defense (Reed 1983, 58). However, in the U.S.-Japan Meeting records, the phrase "free ride in basic research" did not appear. It was probably coined through Japan's interpretation.

In the 1980s, in becoming a major economic power after rapid economic growth, Japan was satisfied with the "catch-up" to advanced countries. In 1987, the term *"international* state" was repeatedly emphasized, especially by Prime Minister Nakasone Yasuhiro, which received a special prize in the 1987 New Words and Buzzwords Award. Japan also aimed at becoming a front-runner country in areas other than the economy. However, the term "international" in Japanese policy at the time was a convenient buzzword; the meaning was inconsistent and lacked concreteness. The phrase "free ride in basic research" was also well adapted to Japan's inferiority complex toward Western countries, coupled with the interests of the Japanese industry and universities to promote basic research.

In the late 1970s, Japan and the United States, respectively, began to build a national innovation system as part of their economic policies and promoted science and technology policies for R&D. The science policies were based on the linear model from basic to applied research and development. Industry-academia collaboration reduced basic research at universities in the United States. However, in Japan, the collaboration was consistent with the interests of universities since the national industrial policy and the United States demanded Japan's active global role in basic research (Kobayashi 1998, 53, 60). The promotion of basic research was consistent with industrial and governmental interests because of its focus on *oriented basic research*. Engineering professors sharing these interests also actively participated in the promotion.[13] The "free ride in basic research" criticism as external pressure from overseas led Japan to enhance research at universities.[14] Basic research promotion led to the enactment of the Basic Act on Science and Technology in 1995.

ENGINEERING EDUCATION REFORM FOR KNOWLEDGE-BASED ECONOMY

The university research institutes and engineering education reforms were promoted following engineering-oriented science and technology policy for R&D through industry-academia collaboration. These reforms led to university-wide reforms in the 1990s. This section first discusses the case of the Research Center for Advanced Science and Technology (RCAST) at the University of Tokyo, established in 1987. Second, it discusses the nationwide educational reform in the mid-1980s and its development into engineering education reform.[15] In the 1980s, various research institutes were founded: the Graduate University for Advanced Studies (in 1988), the Advanced Institutes of Science and Technology (in 1990, Ishikawa, and 1991, Nara), joint research centers of national universities (three in 1987, five in 1988), and endowed chairs. RCAST is the representative case.

The Institute of Space and Aeronautical Science (ISAS), the predecessor of RCAST, was divided into the MOE/ISAS and the Institute of Interdisciplinary Research at the University of Tokyo's Faculty of Engineering; the latter was planned to be abolished in 1988. Furthermore, to resolve the abolition and facility use problem progressively, the preparatory committee of RCAST was organized in 1986. Inose Hiroshi, dean of the Faculty, took the initiative to formulate the plan (RCAST 2007, 17; Nano 1991, 68). The objective of RCAST was based on an *interdisciplinary approach*, *mobility*, *international perspective*, and *openness* stipulated in Article 2 of the Rules for RCAST. These four founding principles were repeatedly mentioned as *mottoes* that characterized RCAST. For example, *mobility* was realized by the "reflux" of the personnel affairs and humanities-sciences combination, and *openness* was realized by recruiting several visiting professors (RCAST 2007, 77, 226; Mikuriya 2008, 42–75). Inose stayed at Bell Telephone Laboratories from 1956 to 1958 to develop the basic idea of his major achievement (the time-slot interchange system)[16] and saw the dynamic research environment as an ideal to reach (Inose 1990, 22–29; Inose and Murakami 1992, 32–54).[17]

Among these principles, *interdisciplinary approach* and *mobility* were issues in the transition from elitist *ivory tower* higher education to education available to the general public and were noted through the college riots in the 1960s. However, *international perspective* and *openness* were issues following the 1980s and were noted as structural problems in Japan in the U.S.-Japan talks.[18] RCAST was expected to be a pioneer in making the University of Tokyo an internationally open university, thereby responding to international criticisms of symmetrical access and the "free ride in basic research." In particular, no large-scale private donation system existed in Japan's national universities. Thus, Japanese companies donated significant research grants to American universities, coupled with the unilateral flow of researchers from Japan to the United States.

Hence, the National School Establishment Act was amended in May 1987 to resolve these problems. Meanwhile, RCAST introduced Japan's first endowed chairs from private companies, and eight visiting professors with a fixed term were invited via the donation from NTT, NEC, CSK, and Nippon Steel Corporation (Okoshi 1992, 1–39; Public Relations Office, University of Tokyo 1987, 5).

The introduction of the endowed chairs saw considerable criticism that it would damage academic freedom due to industry-university collusion. Moreover, fixed-term visiting professors were not obliged to attend faculty meetings nor did they retain the rights to vote on personnel and budget. The system was criticized as a "cancer cell," shaking the university's autonomy and professorship (University of Tokyo Faculty and Staff Union 1987). Inose explained these criticisms as follows:

> If anything, they are still stuck on the ethical aspect. There is a principle of segregation of public officers from private companies, and the professors of national universities and the MITI officials are the same national public employees. It is considered very shameful for the professors at a national university, who are government employees, to receive a salary from a private company for collaboration in R&D, just as it becomes corruption for the MITI officials to benefit a private company. Furthermore, there is an ideological argument. Some believe that supporting the research of private companies must not be permitted due to its counter-revolutionary nature because capitalists are essentially evil in general. It seems to become a criticism of industry-academia collusion. (Inose and Murakami 1992, 103)

Thus, RCAST strived to become an organization without conflict of interests by asking companies to donate without any request in principle and initially inviting all visiting professors from abroad (Inose 1990, 82–85; 1994, 19–20).

Mori Wataru, president of the University of Tokyo at that time, was the first president from the Faculty of Medicine since the University of Tokyo Riots in the late 1960s. He steadily conducted the campus reforms to liquidate the riots. Medical students' indefinite strike initiated the riots over the unreasonable intern system in January 1968. Mori evaluated the riots very negatively, stating that much had been lost and very little had been gained. He then reconstructed the Yasuda Auditorium, a symbol of the university and the riots, implemented regular environmental improvement weeks, including removing many offensive signboards by students, and called on university faculty and staff to work honestly as public servants (Mori 1989, 182–187).

They were "ordinary" policies, as Mori stated, but he regarded them as a duty to terminate the bad habits related to the riots. For Mori, the declining presence of universities in society was attributed to their lost public credibility and other debt via the college riots (Mori 1989, 177–188).

Mori was also concerned about the "free ride" criticism. In his presidential address at the entrance ceremony in 1986, reflecting on the history of the university since the Meiji era and referring to the recent reputation that Japan had become a "first-rank country" and a "rich country," he worried about the current situation of the university:

> However, in reality, the degree to which Japan has participated in and contributed to the development of the world's culture is surprisingly small. Until today, Japanese people have worked on everything with an attitude to try to catch up with advanced countries. As a result, we are accustomed to being cared for by others in terms of culture and cannot imagine the value. We also seem to be ignorant of the preciousness and difficulty of leading in building it, thereby neglecting efforts. Speaking bluntly, we tend not to be ashamed of the free ride in culture but only seek efficiency . . . this is actually a problem. (Mori 1989, 54–55)

After referring to Japan as not making cultural contributions per its national power, he stated: "Thus, perhaps, many foreigners are starting to feel that we are repugnant, assuming Japan to be a free-ride nation in its culture" (Mori 1989, 55–56). Mori, president of the University of Tokyo, regarded free riding as a problem of Japan and called for improvement. Although the emphasis on basic research began circa 1980 before the U.S.-Japan talks, it was closely related to the inferiority complex against Western culture, which nurtured modern science. Therefore, Japan reacted nervously to the reputation of taking a "free ride," and the "free ride in basic research" criticism became very persuasive.

In addition to improving universities' research environment to address international criticism, comprehensive educational reforms were promoted for domestic issues. In 1984, the Provisional Council on Education Reform (Rinji Kyōiku Shingikai) was established to reform the Japanese education system as an organization to advise Prime Minister Nakasone, who employed the slogan of "settling all accounts of postwar political issues" (*sengo seiji no sōkessan*). Although the postwar democratic education system had improved the educational standard and education continuance rate across the country, competition for entrance examinations was heated due to the excessive emphasis on educational background. Moreover, frequent occurrences of bullying (*ijime*), refusal to attend school, and school violence had become a social problem. From 1985 to 1987, the Provisional Council reported that autonomy and creativity, as well the Japanese identity, were not sufficiently developed in elementary and secondary education due to the excessive emphasis on rote learning in the uniform and inflexible education system. Furthermore, universities had also not risen to adequate international levels. Hence, the Provisional Council first emphasized the principle of individuality, followed by autonomy and responsibility, fundamentals of character-building, creative thinking and the expressive ability for internationalization and computerization, and transitioning to a lifelong learning system (Provisional Council on Education Reform 1988, 7–23). The Provisional Council also presented specific policies to universities, including the development of a lifelong learning system, promotion of basic research (e.g., expansion of the postdoc system), and a fundamental readjustment of the Standards for Establishment of Universities (SEU) (Provisional Council on Education Reform 1988, 283–291).

Given these policies, the MOE University Council was established in 1987 to submit a report to the MOE minister. Based on the report, the SEU was generalized in 1991, allowing each university to develop autonomously unique educational and research systems. Through this generalization, systematic separations such as those between general education and specialized education subjects were abolished or made more flexible. As the requirements were relaxed, self-examination and self-evaluation systems for educational and research activities were introduced. An American accreditation system was cited as a peer review model, which could not be realized by the postwar JUAA, established under the GHQ/SCAP initiative. Nonetheless, most universities addressed the SEU's generalization stereotypically and inflexibly and reduced general education, a symbol of the postwar democratic educational reform, while expanding specialized education.

Based on the university education policy, the MOE Higher Education Bureau organized the Investigation Collaborator's Council on the Promotion of Engineering Education (Kōgaku Kyōiku no Shinkō ni Kansuru Chōsa Kyōryokusha Kaigi) in October 1988 and published a report, *Engineering Education in the Transitional Phase* (*Henkaku ki no kōgaku kyōiku*), in December 1989. Until then, the MOE had not intervened in the course contents, given universities' autonomy, and restricted its authority to the coordination of university budget requests (Kusahara Katsuhide, interview by the author, October 19, 2012). In contrast, this deliberation considered the contents. While appreciating that Japanese engineering education was successful in training and supplying high-quality R&D human resources, they illustrated the following five aims: (1) eliminating an imbalance between the increasing demand for engineering human resources and its supply; (2) improving the flexibility of the education system to meet the increasing need for interdisciplinary approaches; (3) improving graduate schools to strengthen the foundation of basic research; (4) developing human resources and lifelong learning to promote innovation; (5) responding to the practices of international education systems, such as admission, degree conferment, and teacher employment. Advocating the SEU's generalization, the 1989 MOE report requested the realization of an education system under "free competition based on evaluation" rather than "control by standards" principles to address the transition to a contract society (Investigation Collaborator's Council on the Promotion of Engineering Education 1989, 1–12; Ichikawa et al. 1990, 5). At the time, the *White Paper on Science and Technology 1989* noted the science and engineering students avoiding the manufacturing industry. In contrast to the increasing demand for engineering human resources, the decline in students' aspirations to choose engineering education and work became a social problem. Japan was in the midst of the so-called bubble economy boom, and the pampering and showiness of humanities-related companies, such as finance, had spurred students' low interest in engineering.

Despite noting the need for external assessment, the 1989 MOE report was only concerned with domestic issues and did not expect any reform in response to the international trend in engineering education.[19] Regarding engineering qualification and education, each country's restructuring and international multilateral agreements on the mutual recognition of substantial equivalence were progressed in the 1980s to respond to the transformation of the global industrial structure. In the United States, the Accreditation Board for Engineering and Technology (ABET), renamed from the ECPD in 1980, signed the Washington Accord with engineering societies in other English-speaking countries in 1989. In Europe, a registration system for the Européenne Ingénieur (Eur Ing) was launched in 1987 under the Fédération Européenne d'Associations Nationales d'Ingénieurs (FEANI). The United Kingdom founded the Engineering Council (EngC) in 1981 and enacted the Education Reform Act in 1988. Although internationalization in Japan focused on how to increase the number of international students, universities did not feel the need to be compatible with other countries. The 1989 MOE report emphasized the introduction of a self-evaluation system; however, it came from a common domestic objective of the MOE University Council to cut into taboos to reduce the stagnation of universities (Kusahara 2012, 17; Kusahara, interview by the author, October 19, 2012). The MOE-led proposal for engineering education reform ended with the publication of the 1989 MOE report, and the reform was left to the policies of each university along with the SEU's generalization.[20]

The 1989 MOE report emphasized the introduction of a well-balanced curriculum, which included humanities subjects, such as philosophy and ethics, and the introduction of classes focusing on the relationship between technology and society, including *technology studies* and *the history of technology*.

It further suggested that a wide range of *techno-literacy* (kōgaku kyōyō) was also necessary for non-engineering students (Investigation Collaborator's Council on the Promotion of Engineering Education 1989, 17–18; Ichikawa et al. 1990, 10–13). In 1989, ethics education was not recognized as important.

In contrast, the Engineering Academy of Japan (EAJ), founded in 1987 by industry-academia engineering leaders, was actively interested in the issues of accreditation and internationalization. The EAJ organized the Engineering Education Committee, chaired by Imai Kaneichiro, director of the Japan Technology Transfer Association (JTTAS) and addressed the engineering education issues as a joint committee with the SCJ's Section 5 (on engineering), the Basic Engineering Research Liaison Committee (Kiso Kōgaku Kenkyū Renraku Iinkai). In 1989, the committee conducted tours in the United States and European countries and confirmed that a qualitative shift in engineering education occurred to enhance international competitiveness per the changes in industrial structures and the active development of accreditation institutions such as ABET (Hirooka 1990). After the investigation, the EAJ and the SCJ published the joint proposal, *Development of Human Resources and Systems to Support Tomorrow* (*Asu o sasaeru jinzai ikusei to taisei seibi*), in 1990. Furthermore, the SCJ published the proposal, *The Problems and Responses to Engineering Education* (*Kōgaku kyōiku ni kansuru shomondai to taiō*), in 1991. In the 1990 proposal, they called for drastic measures, fearing that Japan would be isolated from the common international base of engineering education and the licensing of engineers, such as the Eur Ing and the Washington Accord, due to the lack of international perspective (EAJ Engineering Education Committee 1990). Okamura Sogo, who was then the director of the SCJ's Section 5, quoted the specific suggestions in the Report of the Engineering Education Mission to Japan in 1951, criticizing as follows: "These are opinions that were stated about forty years ago, but they still hold true today. In other words, engineering education has not been reformed at all even after forty years" (Okamura 1991, 8).

During that time, Harada Kosaku, the JSEE secretary-general and a member of the 1989 MOE report, attended the International Symposium on Higher Engineering Education held in Zhejiang, China, in 1990. He reported his findings on ABET in the journal of the JSEE. He argued that internationally acceptable degrees and the PE license are essential for the age of internationalization (Harada 1991, 44). The book on the history of JABEE evaluated this as the beginning of the research project to establish JABEE (JABEE 2003, 1, 85, 91).[21] However, it was six years later that the JSEE established its preparation committee.

In the early 1990s, the focus of the MOE Higher Education Bureau, the JSEE, and industry interest in engineering education shifted to the *refresh education* for engineers (MOE 1994, 7–21, 33–41; Kajino 1996, 19–20).[22]

Harada also served as the secretariat of the JSEE Investigation Committee on Refresh Education in Engineering, commissioned by the MOE, from 1992 to 1993. In refresh education, as well as the 1989 MOE report, the expected subjects of the humanities and social sciences were *the history of science and technology* and *science and technology studies* for introduction to engineering. Ethics was only considered a subject for liberal arts education alongside philosophy (Research Committee on Refresh Education in Engineering 1993).

Despite international progress in engineering education accreditation, other countries did not request its introduction to Japan. Moreover, engineering ethics education did not become a topic in international talks. For example, at the Third U.S.-Japan Meeting in 1991, the U.S.-Japan Joint Task Force on Engineering Education was launched, co-chaired by Okamura and Mildred Dresselhaus at MIT. In the working group from 1992 to 1998, Committee 149 and the National Research Council (NRC) discussed engineering education for training *global engineers* and exploring major similarities, differences, and trends in all educational stages (K–12, undergraduate, graduate, and continuing education). The recruitment of young people, especially women and minority groups, was discussed as an issue common to the United States and Japan. Moreover, a simpler and diverse university entrance system, a rigorous undergraduate curriculum, and the expansion of industry-university cooperation were discussed as the challenge for Japan. Regarding the education of global engineers, discussions touched on cultural skills, knowledge of the business and engineering cultures of counterpart countries, the accreditation system of ABET, and the Washington Accord but not particularly on engineering ethics (NRC 1999; JSPS Committee 149 2000, 54–56, 133–139).[23]

The worldwide reforms of engineering education, beginning in the 1980s, can be understood as a restructuring of the knowledge-based economy, which was described by the OECD as "economies which are directly based on the production, distribution and use of knowledge and information" (OECD 1996, 7), focusing on the functions of human capital and technology in economic growth through the high-technology industry. This idea was similar to Peter Drucker's "knowledge economy," Daniel Bell's "post-industrial society," and Alvin Toffler's "the third wave." It gained worldwide popularity through patent policies in the United States and other countries.

In the 1980s, while the Nakasone Cabinet promoted "settling all accounts of postwar political issues," academic research and education were incorporated into market-based industrial policies through economic talks with the United States and European countries. Until then, universities were independent of economic activities but could not keep up with the increasing costs of aged research facilities. Thus, engineering professors, whose works were related to industrial policy, actively cooperated with the reform. Moreover,

Japan's industry and government appreciated it. Although, in theory, it should be targeted only at engineering education, it induced nationwide university education reforms in the 1990s by introducing educational assessment with the SEU's generalization. Meanwhile, international engineering qualification and education frameworks did not become a major issue until the mid-1990s.

INCREASING SOCIAL DEMANDS FOR APPLIED ETHICS

In Japan in the 1970s and 1980s, in contrast with the lack of attention to engineering ethics, new ethical issues emerged with science and technological progress. Thus, applied ethics in environmental issues, life sciences, and information technology were developed. Environmental ethics was noted by Rachel Carson's accusations in *Silent Spring* in 1962 about the severe effects of synthetic pesticides on ecosystems. It gave way to criticisms of anthropocentrism and Judeo-Christian traditions in Western civilization. In 1972, the Declaration of the United Nations Conference on the Human Environment (Stockholm Declaration) was announced, and the Club of Rome, in *The Limits of Growth*, warned of the decline in food production and natural resources due to industrialization, pollution, and population expansion. Furthermore, in 1973, Ernst F. Schumacher criticized overconsumption in economic activities and defended *appropriate technology* in his book, *Small Is Beautiful*. In the same year, the first oil crisis occurred and evidenced the need to change social values around energy consumption. In the 1980s, acid rain and ozone depletion became serious, and the Chernobyl nuclear power plant incident caused transnational environmental pollution in 1986.

Meanwhile, medical ethics were developed based on the philosophy of civil movements, such as the civil rights movement, the consumer movement, and the women's liberation movement in the United States in the 1960s. In 1962, it was reported that the Admissions and Policies Committee of the Seattle Artificial Kidney Center (the so-called "God Committee") had screened patients to deploy a limited number of medical devices for hemodialysis. Supported by the WMA Declaration of Helsinki on medical practice in 1964, Henry K. Beecher at the Harvard Medical School disclosed in 1966 that clinical research was being conducted like human experiments in the United States. It emphasized the importance of ensuring informed consent and self-determination. Moreover, in 1972, the Tuskegee study incident occurred, in which syphilis patients of African descent were treated as experimental subjects. As these became social issues, the Statement on a Patient's Bill of Rights was adopted in 1973 against physicians' unilateral paternalism, the

National Research Act was enacted in 1974, and ethical principles for medical experiments were established.

Van R. Potter first coined the term *bioethics* in 1971, which was a general ecological ethics for human survival that evolved from environmental ethics. Meanwhile, new technologies, such as recombinant DNA, genetic disease screening, organ transplantation after brain death, and artificial insemination, were introduced in the 1970s. These enabled artificial manipulation of phenomena with which fundamental human life values were concerned, such as changing the line between life and death and human rights and privacy rights for mothers and aborted fetuses. Thus, bioethics became a social issue with a different meaning than the original. In particular, in 1975, Karen A. Quinlan's case fought for the right to die in a persistent vegetative state. Furthermore, the Asilomar Conference on Recombinant DNA was held.

In Japan, the Hippocratic Oath was noted as a code of medical ethics, and the JMA established its code of ethics in 1951. In contrast with the focus on physicians' professionalism, the new medical ethics evolved to focus on patients in the late 1960s. Discussions in Japan evolved independently of those in the United States. For example, the Wada Juro heart transplant case came to represent a major social problem in 1968. The Mitsubishi Kagaku Institute of Life Sciences, established in 1971, also addressed ethical issues, advocating *life sciences* as an integrated framework that incorporates the pollution issues caused by economic growth. In the 1970s, the Blue Grass Group (Aoi Shiba no Kai), part of the women's liberation movement, campaigned for disabled people's right to live, given prenatal diagnosis.

In becoming an international issue, bioethics was imported from the United States to Japan, thereby beginning on a large scale in Japan.[24] The Ministry of Health and Welfare (MOHW) and the Diet established councils on bioethics in 1983. With Prime Minister Nakasone's proposal at the Ninth G7 Summit of the year, the Conference on Life Sciences and Mankind was held in 1984, inviting Western researchers to Japan. In the same year, based on the SCJ's report, the Basic Plan for Research and Development of Leading and Fundamental Technologies in Life Sciences was launched as a national policy, and the industrialization of life sciences progressed. Religious debates, such as those on Buddhist ideas of life and death and Christian ideas of the restoration of humanity, also developed. The Japan Association for Bioethics was established in 1988 (Kagawa and Komatsu 2014).

Bioethics was integrated with environmental ethics and further developed as comprehensive ethics of technology or *techno-ethics* for coordinating ethical, religious, legal, political, and other issues in a multicultural society. For instance, in 1984, the Japanese Society for Ethics chose the theme of "technology and ethics" for the annual conference and published the symposium outcomes, "Fundamental Problems of Modern Technology," as an edited volume in 1985. It highlighted various cultural perspectives with reference to *The Structure of Scientific Revolutions* by Thomas Kuhn and emphasized the need to reexamine the ethics that had been an implicit premise in the twentieth century (Japanese Society for Ethics 1985, 1–2). Imamichi Tomonobu proposed the *eco-ethica* studies, which interpreted science and technology not as mere tools but as ecological environments causing changes in social values.

Some universities began to work systematically on technology ethics research. In 1985, the Nanzan University Institute for Social Ethics, a Catholic university, began publishing a bulletin, *Technology and Ethics in the Modern World*. Johannes Hirschmeier, university president, established the institute in 1980 as the Nanzan Institute for Economic Ethics to focus on the negative aspects of economic development from a combined perspective of traditional Japanese ethics and Christian ethics. The following year, the research subjects were extended to include the law, and the name was changed to a more comprehensive one (Nanzan University Institute for Social Ethics 2006, 217). Anan Seiichi, who specialized in natural law, served as the director from 1983, and set up the continuous theme of *technology and ethics* from 1985, holding research meetings by 1992 on legal philosophy (five meetings), economic philosophy (eleven), and bioethics (fourteen) (Nanzan University Institute for Social Ethics 2006, 218–219). The number of meetings shows that bioethics was of the greatest concern at that time. Meanwhile, Anan (1987) also paid attention to engineering ethics and discussed the NSPE code of ethics with reference to Mike W. Martin and Roland Schinzinger's textbook (1983). Furthermore, the necessity of information ethics and its professional ethics were argued for (Maekawa 1987, 48–51). However, Anan's research was generalized to *homo utens technologia*, an essentialist speculation on human beings' utilization of technology (Anan 1991), and the title of the bulletin was changed to the *Nanzan Forum for Social Ethics* in 1992 to eschew the limitation to technology.[25]

In the late 1980s, ethical issues of advanced science and technology became widely recognized. Murakami Yoichiro, professor at RCAST, reduced science and technology ethics to professional ethics existing at an intersection with engineering education reform. He was transferred from the Department of History of Science at the same university to the Science and Technology Ethics (*kagaku gijutsu rinri*) Research Category of RCAST, newly established in 1988. Moreover, from 1993 to 1995, he served as the director of RCAST, despite being from the humanities.[26] Given his experience at RCAST, he published a book titled *What is the Scientist?* (*Kagakusha towa nanika*), which questioned the professional ethics of individual scientists (Murakami 1994a, 182–186). It was intended for scientists, not engineers, but described science and engineering in general, recognizing that "scientist" had not yet become a fully realized profession. Although the issue was already known, in the 1980 SCJ Charter at least, he wrote it as a contemporary, easy-to-understand book that focused more on the ethics of individual experts. Thus, his book became very popular in subsequent discussions of engineering ethics. He also delivered an invited lecture at the annual convention of the Information Processing Society of Japan (IPSJ) in March 1995, which became the first Japanese academic society to establish a code of ethics since the 1990s, as described in the next section.[27] He has been active in numerous lectures, writings, and project management on science and technology ethics since the mid-1990s.

In his book, Murakami argued that scientists were not only professionals but also entrepreneurs, created in the nineteenth century, whose behavioral principle was nothing but *free competition* from inception; therefore, ethics were necessary (Murakami 1994a, 13–14).[28] Criticizing not only their attitude to always move forward with "something new–ism" like "automobiles without brakes" but also their vagueness of social responsibility and obligations to the broader community, he emphasized the importance of accountability and transparency to society (Murakami 1994a, 104–106, 158–181).

Murakami began his discussion by introducing the criticism by Karaki Junzo (1980), a literary critic. Karaki particularly urged Japanese physicists to be aware of their being complicit in the *original sin* and perform their social responsibility because they were self-servingly discussing the atomic bombings and postwar nuclear weapons development. Murakami was shocked when reading the pre-published draft of Karaki's book. Moreover, when Murakami translated and published Erwin Chargaff's autobiographical essay, *Heraclitean Fire: Sketches from a Life Before Nature*, Murakami was strongly influenced by Chargaff's warning that the scientific community was autonomously and automatically reproducing itself and bloating. Even the biologist Chargaff blamed himself for being as guilty as the scientists who developed the atomic bomb (Murakami 1994a, 14–17; Murakami, interview by the author, February 17, 2014; Chargaff 1980, 259–260). Murakami is Catholic; he denied any influence of religion on his academic research (Murakami 2016, 68–69). However, it seems to explain his strong sensitivity to the concept of sin.

Furthermore, in the 1980s, scientific misconduct attracted attention in the United States. The *Betrayers of the Truth* by William Broad and Nicholas Wade was published in 1982 and translated into Japanese in 1988. The U.S. Public Health Service established the Office of Scientific Integrity in the National Institutes of Health. Moreover, it established the Office of Scientific Integrity Review in the Office of the Assistant Secretary for Health in the Department of Health and Human Services in 1989. These establishments were consolidated into the Office of Research Integrity in 1992. Robert I. Bell published *Impure Science: Fraud, Compromise, and Political Influence in Scientific Research* in 1992, which was also translated and published in Japan in 1994.

Meanwhile, Murakami's argument largely related to the development of science and technology policy in the 1980s. RCAST was established per the national policy to strengthen industrial competitiveness in the 1980s, and Murakami was required to tackle ethical issues within its philosophy. Furthermore, he became an editorial board member of the international academic journal, *Sociology of the Sciences*, in 1992, in which role he was influenced by the discussion of technological innovation in modern society (Murakami, interview by the author, February 17, 2014). Its co-editor, Björn Wittrock, published a collection of essays, *Science as a Commodity*, with Michael Gibbons. They analyzed the current situation in which scientific community openness as a traditional value was changing due to commoditization. However, they argued that, historically, the professional and autonomous activities of scientists had been supported by commoditization. Hence, they discussed the importance of socially responsible science and technology policy (Wittrock and Gibbons 1985). Murakami once wrote in 1979 that scientists were only "normal human beings," and "noble personality" or "special ethics" could not be expected merely from being a scientist (Murakami 1979, 201). This understanding was similar to that in the 1990s, but he reinterpreted the scientist's principles of behavior per the introduction of industrial policy and market principles in science and technology at the time.

While interest in science and technology ethics had increased, demand in corporate ethics had been increasing since the late 1980s, promoting its systematic adoption. After the Toshiba Machine–Kongsberg violation of the Coordinating Committee for Multilateral Export Control (CoCom) in April 1987 through the export of machine tools to the Soviet Union (considered to have contributed to the quietness of the submarine propeller), formulating a corporate ethics program was required to realize *compliance*, a new concept in Japan. Meanwhile, Tateho Chemical Industries (September 1987) and Nippon Steel and Sankyo Seiki (July 1988) were allegedly involved in stock insider trading. Furthermore, 1988 saw a large-scale bribery scandal, where Recruit Co. transferred unlisted shares to many politicians. Thus, the Securities and Exchange Act was amended in 1989 to regulate insider trading, and corporate governance was restructured, though insufficient with the 1981 amendment of the Commercial Code (Unayama and Ogura 1996, 79–92). In response, Keidanren established a Corporate Ethics Committee in 1989 to survey and implement corporate ethics and adopted the Charter of Corporate Behavior in 1991. Moreover, in the U.S.-Japan Structural Impediments Initiative (SII) Talks in 1990, the United States urged Japanese companies to develop antitrust compliance programs. The Japan Fair Trade

Commission published the Guidelines on Antitrust Law in 1991, further increasing interest in compliance.

The emphasis on corporate ethics in the 1980s was due to economic and global factors, such as the promotion of deregulation per the idea of a liberal economy, the transition to the knowledge-based economy, compliance with global standards, and the escalation of environmental problems. Moreover, in the early 1990s, the economic bubble burst, the industrial environment worsened, and corporate scandals and bankruptcies continued. Thus, corporate distrust increased in society. The general public began to shift from corporate-centric to social life-centric values. Hence, preventing sentiments such as "I thought it was bad as an individual, but I had to do it for my company as a business" became a vital corporate-ethic goal (Mizutani 1995, 1–19). The 1993 collapse of the postwar Liberal Democratic Party administration, incumbent since 1955, was symbolic of the significant social change in Japan.

As described above, various applied ethics, such as environmental concerns, life sciences, and business matters, attracted public attention by the early 1990s. Although engineering ethics was rarely noticed, the Nanzan University Institute for Social Ethics, for example, introduced American engineering ethics. The importance of moral education was repeatedly noted during the prewar period and after the war. Regarding engineers, besides the IEEJ and JSCE cases, Nikkeiren also stated the following in the 1956 proposal on engineering education: "Needless to say, the goal of school education is not only to provide knowledge but also to build personality, so it is extremely important to thoroughly implement engineering ethics at engineering colleges" (Nikkeiren 1956). However, in Japan, engineering ethics education was not required as specific educational content until the 1990s. The fact that professional ethics education had not been implemented as part of a particular curriculum for engineering education is a different issue than the actual ethics of Japanese engineers. In Japan, engineering ethics education was conducted implicitly in daily education communication and was no longer required by society. Meanwhile, the popularization of applied ethics until the 1990s furnished social preparation to accept the introduction of individualistic engineering ethics. When external pressure to meet global standards was exerted, engineering ethics was quickly introduced. The effects of globalization, especially in information ethics, are discussed in the next section.

GLOBAL PRESSURE TO ESTABLISH A CODE OF ETHICS

In May 1996, the IPSJ code of ethics was adopted as a forerunner to that established by the academic engineering societies in the late 1990s. The IPSJ

Code of Ethics Investigation Committee for the arrangement was established in July 1995, chaired by Nawa Kotaro, professor at the Faculty of Law, Niigata University. The committee was established because "the increasing influence of information processing technology on society made the professional ethics of experts involved in this technology more important," and "as international exchange became more popular, inquiries began to come from overseas regarding the code of ethics of the IPSJ" (Nawa and Yoneda 1996, 702).

Regarding the first reason, Nawa (1996) stated that information technology experts addressed the personal information of various nonprofessional users and responded to various social values as society matured. He also maintained that the IPSJ should establish a code of ethics, just as socially influential professionals, such as physicians, architects, and lawyers, were required to have high ethical standards. However, this was not a direct reason for the IPSJ's enactment. Historically, information technology is supported by nonprofessional users. Thus, the IPSJ committee gave less consideration to the concept of *profession*. Furthermore, it did not pay attention to professional education (Nawa, interview by the author, August 22, 2012).

Regarding the second reason, the enactment of the IPSJ's code of ethics was directly influenced by the recommendation from the International Federation for Information Processing (IFIP). IFIP Technical Committee 9 (TC9, dealing with the relationship between computers and society) submitted a preliminary code of ethics in 1988 and held a panel discussion on the validity of a universal code of ethics, "Ethics of Computing: Information Technology and Responsibility," at the Twelfth World Computer Congress in September 1992. After this discussion, the Task Group on Ethics was organized and decided in 1994 to encourage each member-society to formulate a code of ethics using the universal code of ethics as a guideline (Lee and Berleur 1994, 103; Ozeki et al. 1995, 1078–1079).[29] The first issue entrusted by the predecessor to Yoneda Eiichi, who became the IPSJ director for international affairs in May 1993, was the establishment of the code of ethics. His predecessor had refrained from acting because "in Japan, it is meaningless to establish such a thing" (IPSJ 1997, 60). Later, Yoneda offered a criticism that "the dependent nature of Japanese society on the external pressure appears" in the fact that not the director for general affairs (as it should have been) but the director for international affairs was in charge of the issue (Yoneda 1997, 10).

Furthermore, in October 1993, the IFIP task group sent "the last call" with a strict stipulation to the IPSJ, requesting the information on the existence of the IPSJ's code of ethics or the activities for its enactment in Japan. In response to the "demand letter, which may be called a threatening letter," the IPSJ returned an English translation of the code of the Japan Information

Technology Service Industry Association (JISA). However, its content was not the individual professional ethics that IFIP required but was the ethical standards each member company should follow as an organization. In June 1994, they further received a letter of recommendation from the IFIP president. Thus, at the end of July, the IPSJ board decided to organize the Code of Ethics Investigation Committee not on the international committee but the entire society (Yoneda 1997, 11; IPSJ 1997, 60–62).[30]

As a result of discussions among the IPSJ president, the IPSJ secretary-general, and Yoneda, Nawa, who was familiar with the legal and computing aspects, was requested to be the chairperson (IPSJ 1997, 62). Although Nawa belonged to the Faculty of Law and contributed to the IPSJ Magazine, he was not a code of ethics expert. Therefore, he felt that he "lapsed into a position to deal with a strange thing, *ethics*" (Nawa and Otani 2001, 148).[31] Nawa and Yoneda first selected energetic members to gain societal trust and selected external experts (e.g., ethics and legal experts and a journalist) to reduce any bias. They launched the Code of Ethics Investigation Committee with ten members in July 1995. Yoneda served as the secretary of the committee (IPSJ 1997, 62–63; Nawa and Otani 2001, 149).[32]

The IPSJ committee modeled themselves on the Association for Computing Machinery (ACM) regarding inclusiveness after examining the codes of the JSCE, the JISA, the ACM, the IEEE, and the Computer Professionals for Social Responsibility (IPSJ 1997, 65–66; Nawa and Otani 2001, 150). The code was drafted in November via meetings and e-mail discussions. After a symposium at the annual convention and gathering opinions, the IPSJ approved the code of ethics at the general meeting in May 1996 (Nawa and Yoneda 1996), enacting it in less than a year since the committee's establishment. The enactment was so smooth that the IPSJ secretary-general gave praise: "It is unusual for a committee to have such a lively discussion and to have an early output" (Yoneda 1997, 12–13).

Although the establishment of the IPSJ's code of ethics was promptly pushed forward by external pressure, prior discussions on information ethics had also been held at the Institute of Electronics, Information, and Communication Engineering (IEICE). It recognized the necessity of information ethics circa the spring of 1991 and launched the Forum on Advanced Communication Ethics (FACE) in 1993 to hold research meetings and panel discussions. However, the discussion theme was not professional ethics but information ethics in general. Nonetheless, from the report of Kurokawa Tsuneo at Kogakuin University, who participated in IFIP in 1994, they knew that Japan was the only advanced country without a code of ethics; they understood the necessity (Kasahara 1995; Tsujii 1995). In 1995, the IPSJ committee formation year, the FACE was upgraded from the Type 3 study group "to search for new fields close to society and investigate future research themes" to a permanent Type 1 study group "to announce and discuss the investigation results in the core fields of the society, and promote the development and diffusion of knowledge and technologies in the field of its responsibility." Meanwhile, the Code of Information and Communication Ethics Drafting Working Group was established. Finally, the Charter of Ethics was enacted in July 1998. The IEICE/FACE was launched earlier than the IPSJ committee but had more extensive discussions on information ethics. The IPSJ and IEICE committees were established in 1995, but the IPSJ was exposed to immense external pressure as Japan's representative society of information science. The IEICE, which was under less similar pressure, then discussed the issues more carefully (IEICE/FACE 1997; 1999).[33]

Although the necessity of information ethics was generally recognized, the code of ethics was recognized as belonging to a foreign culture. Nawa stated that the establishment was a matter of external pressure for the IPSJ. Thus, it was hesitant to create a code of ethics and, in contrast to its high sensitivity to foreign technical information, was insensitive to international affairs, lagging significantly in ethics (Nawa and Otani 2001, 149; Nawa 1999, 1121). Without the pronounced external pressure from the global standard, there may be no need to introduce the strange culture into Japan, and the response to the culture would be postponed with hesitation. While Yoneda stated that the external pressure increased circa the summer of 1994, he also stated that the series of incidents on safety and ethics in 1995 (e.g., the Great Hanshin-Awaji Earthquake and Aum Shinrikyo cult's sarin attack on the Tokyo subway) accelerated the enactment of the code of ethics (Yoneda 1997, 10). However, these incidents came after the decision to establish the IPSJ committee and were not the direct cause of the enactment. Nevertheless, it was repeatedly emphasized when the introduction of engineering ethics was promoted in Japan. Chapter 7 considers the influence of the 1995 incidents.

When Murakami published *What Is the Scientist?* in 1994, and the IPSJ established the code of ethics in 1996, attention to ethics for individual engineers increased in Japanese society. The former was influenced by Murakami's experience at RCAST, and the latter by the IFIP's recommendation for enactment. Both were based on transforming the global industrial structure in the 1980s toward a knowledge-based economy; however, neither may have been fully aware of the economic trend. As the former was an extension of Murakami's sociological research since the 1980s, and the latter, like the JCEA case, was restricted to the introduction of a code of ethics, neither envisioned professional ethics education. A further impact of economic globalization was necessary for Japanese society to be steered toward adopting American-style engineering ethics education. Chapters 6 and 7 discuss the next stage of the process.

NOTES

1. The establishment of a code of ethics for scientists was also promoted by the data fabrication detected in 1974 when the Japan Analytical Chemistry Research Institute, commissioned by the STA, faked the radioactivity measurements of U.S. nuclear submarines (Fushimi 1980, 21).
2. The WMA adopted the International Code of Medical Ethics in 1949.
3. Kimura also received information that an engineering council was considering a charter for engineering researchers; the JFES may have considered a charter for engineers' equivalent to the SCJ Charter, but details are unknown (Kimura 1979, 44–45).
4. The Japan Teachers' Union (JTU) joined the World Organization of the Teaching Profession (WOTP) in 1951 and adopted the Code of Ethics for Teachers in 1952, as per the 1951 WOTP Assembly role in setting the theme of professional conduct. The JTU's efforts to assert the rights of teachers as laborers gained international support from the ILO and UNESCO's Recommendation Concerning the Status of Teachers but conflicted with the government's policy of demanding their devotion to education. It is also possible to suppose that the IEEJ rejected the proposal because the code of ethics was associated with the communist labor movement.
5. When Inose Hiroshi drafted a report to the government circa 1970 that science and technology would become an international and domestic policy issue, it was rejected as a fabrication because it had no precedent (Inose 1986b, 363).
6. Japan National Railways was privatized to the JR Group, the Nippon Telegraph and Telephone Public Corporation to the NTT Corporation, and the Japan Monopoly Corporation to Japan Tobacco Inc. and the Salt Industry Center.
7. Pure basic research differed from oriented basic research since the former is distinguished as "research for new phenomena or knowledge with unknown potential for application" and the latter is "research to fill a currently unknown gap to serve a particular purpose" (STA 1963).
8. Although the term *basic research* was introduced in policy-making, the term *basic science* was also commonly used in Japan. For example, the 1960 CST's advisory report explained "promotion of basic science" as follows: "basic science and applied science constitute science and are thereby distinguished from technology" (CST 1960, 2). This classification continued in the twenty-first century; for example, the SCJ's 2010 recommendation expanded the meaning of basic science, stating that "the basic science described here refers to the overall intellectual creative activities at universities, etc., including basic and applied research as defined in the OECD Frascati Manual" (SCJ 2010, ii).
9. The MOE Bureau of Science and International Affairs was established in 1974 by reorganizing the University and Academic Affairs Bureau for promoting international exchanges and the planned expansion of higher education. The director-general served as the chairperson of the Japanese National Commission for UNESCO.
10. Okamura and Inose were the SCM members; Okamura was the chief. In 1973, the SCM had already reported on the mobilization of research institutions and

internationalization of universities as basic measures for the promotion of science (SCM 1973, 4).

11. The United States was frustrated with the unilateral flow of students and researchers from Japan to the United States to obtain basic U.S. research results, as well as the trade imbalances and language barriers. In response, Japan argued that it was inevitable that more Japanese researchers visited the United States than American researchers visited Japan because there was a large difference in basic research funding between the two countries.

12. These measures included symmetrical access; acceleration of science communication, such as via workshops; responses to the Newly Industrialized Countries (NICs) in the Pacific Rim; responses to trade issues of advanced technology at international conferences, such as the GATT and the Economic Summit; and establishment of an engineering academy to visualize knowledge. Regarding the engineering academy, Esaki Reona, a Nobel Prize winner, proposed the concept to the government and the SCJ, and the 1980 Industrial Structure Council report also proposed the establishment of an organization modeled on the NAE (RIITI 1980, 95). Although the discussions were not settled easily, an advisory meeting was organized in 1983, and a planning committee was established in 1985. Shortly thereafter, the recommendation to establish an academy was made at the U.S.-Japan Meeting. Finally, the Engineering Academy of Japan (EAJ) was established in April 1987 (Kobayashi 1987, 35–36).

13. Inose advocated a non-purpose-oriented "useless research institute" (Muyōgaku Kenkyūsho), stating: "It is difficult to judge whether truly basic research is useful or not, profitable or not, and the success rate is extremely low" (Inose 1987b, 230–231).

14. This recognition continued afterward. Sawada analyzed Japanese industry-academia collaboration in the late 1990s and early 2000s and explained: "It had an almost nationalistic tone, based on the so-called *linear model*, and sounded pleasant to the professors who followed it" (Sawada 2011, 142).

15. In the MOE, the former was in charge of the Academic and International Affairs Bureau and the latter was in charge of the Higher Education Bureau.

16. A basic technology for Pulse Code Modulation (PCM) transmission and switching in digital telephone networks.

17. Inose stayed in the United States on the Fulbright Grant Program; he was cared for by Koga Issac, a member of the Fulbright Commission (Inose 1990, 25–26).

18. Similar ideas to mobility and openness were expressed in the RCAST principles and reform ideas of the Faculty of Engineering circa 1970 after the college riots. The ideas of internationality, however, were not expressed. Furthermore, openness generally related to the faculty management, not to the research. Other universities assumed the interdisciplinary approach but were limited to engineering. After all, industry-academia collaboration was not proposed either as a mobility or interdisciplinary approach (Faculty of Engineering, University of Tokyo 1970, 1–2). In the 1989 Gourman Report, which rated international universities, the University of Tokyo, the highest ranked university in Japan, was sixty-seventh in the world. This low evaluation relative to the Japan's international competitiveness in technology and economy reminded engineering professors of the need for university reform.

19. As discussed by Theodore Levitt (1983) in *Harvard Business Review*, the United States regarded Japan as an advanced globalized country in the mid-1980s.

20. In 1991, the Eight-University Council of Deans of the Engineering Faculties (Hachi Daigaku Kōgakubuchō Kondankai) voluntarily wrote the proposal to improve graduate schools titled "Mirai o hiraku kōgaku kyōiku" (Engineering Education Opening up the Future) and submitted it to the MOE, the University Council, and Keidanren. This meeting was held by eleven deans of the former Imperial Universities and Tokyo Institute of Technology.

21. Yoshikawa Hiroyuki had participated in the OECD/CSTP since 1987, where the quality assurance of education was an agenda in the analogy with the quality control of the automotive industry (JABEE 2012, 5).

22. The idea of lifelong education became common in the 1960s. The demand for competence development and vocational (re-)training was increasing to respond to technological innovation and computerization, as well as increasing life expectancy and leisure time. Therefore, UNESCO decided to promote *éducation permanente* (permanent education), which harmonized and integrated school education with adult education, at the Third International Committee for the Promotion of Adult Education in 1965. Based on this idea, the French name was translated into English as *lifelong integrated education* and abbreviated to *lifelong education*. Furthermore, the OECD Center for Educational Research and Innovation proposed *recurrent education* in 1973 to circulate working and learning opportunities throughout life. In 1974, the ILO adopted the treaty on paid educational leave and made its recommendation. Per the international trend, the MOE Central Council for Education reported the issue in 1981, and the Provisional Council on Education Reform called for the transition to a lifelong learning system in the 1987 report. This policy aimed to change the school-centered system that was considered to be the cause of the excessive emphasis on educational background. Thus, to address this issue, the MOE reorganized the Social Education Bureau into the Lifelong Learning Bureau in 1988 to promote *recurrent education*, and the MOE Higher Education Bureau promoted *refresh education* for professionals. There is a difference between *recurrent education* and *refresh education*: the former was lifelong education for individuals, including general education; the latter was professional education for groups of professionals.

23. Imai also participated in the working group.

24. The *Encyclopedia of Bioethics*, published in 1978 by Warren T. Reich et al., at the Kennedy Institute of Ethics at Georgetown University, founded in 1971, became a major textbook of bioethics for Japan.

25. The title was changed to include articles on a wide range of topics since it did not impress readers (Nanzan University Institute for Social Ethics 2006, 1).

26. RCAST introduced *science and technology ethics* in 1986, before the establishment (*Nikkei Shimbun* 1986).

27. For the contents of the lectures, see Murakami (1994b, 1996).

28. Murakami argued that "the mid-nineteenth century was a time when the idea of *free competition* in capitalism was believed and implemented in its purest form," and "it deeply penetrated the community of scientific experts" (Murakami 1994a, 49–50). However, since the number of scientists in the nineteenth century was so small that

they formed a visible community, it was doubtful how much the free competition principles worked in the mid-nineteenth century.

29. The report concluded that it did not assert a standard code of ethics, but it presupposed that each society should somehow develop its own code of ethics.

30. At the IPSJ Annual Convention in March 1995, Murakami Yoichiro delivered an invited lecture based on his 1994 book (Murakami 1996).

31. Nawa stated that there was no prospect for the code of ethics at first; he thought it was "fifty-fifty" (Nawa and Otani 2001, 149).

32. Yoneda served as the director for international affairs from May 1993 to May 1995 and, after retiring from the board, was appointed the secretary of the committee (Yoneda 1997, 10).

33. Tutiya Syun at Chiba University was a member of both the IEICE and IPSJ committees. Tutiya promoted the "Construction of Information Ethics" project with Mizutani Masahiko and Ochi Mitsugu as a JSPS Research for the Future Program for five years from 1998. The IEICE/FACE provided the opportunity for the three to meet at the same time (Ochi et al. 2000, 320–323).

Chapter 6

The Globalization of Engineering Qualification and Ethics

INTERNATIONAL EQUIVALENCE BETWEEN GIJUTSUSHI AND PROFESSIONAL ENGINEER

As described in the previous chapter, the progress of globalization in the early 1990s forced the IPSJ to enact a code of ethics ahead of other academic societies. The JCEA had already established codes of ethics in 1951 and 1961, following those enacted by the exemplary American societies and FIDIC international standard. Over time, in addition to the existing code, the JCEA faced another demand for extensive implementation of engineering ethics in the 1990s. In September 1998, volunteers of the Technical Translation Center (TTC, Gijutsu Hon'yaku Center), a JCEA project team, translated and published the engineering ethics textbook of Harris et al. (1995, 1998), *Engineering Ethics: Concepts and Cases*. The JCEA's fiftieth-anniversary book, published by the association in 2001, states that "Japan did not have the concept of *engineering ethics*," and their "voluntary" translation and publication of the American textbook marked the "origin of thinking about ethics education for engineers" (IPEJ 2001b, 221). The systematic approach to ethics in engineering education was a new paradigm for Japan. Although the JCEA had already established ethical codes, which the engineers voluntarily prepared, it never asked for engineering ethics education.

The establishment of the Japanese Gijutsushi system in 1951 was initiated by Tanaka Hiroshi's U.S. report on the American CE profession. Although he found that the American CE business was based on the PE license system, Japanese engineers aimed to introduce only consulting businesses in Japan, not a general engineering qualification like PE. However, the Japanese CE Act in 1957 was extended beyond the American CE business and introduced a unique Japanese system to improve the social status of general high-level engineers. Thus, Japan's Gijutsushi system was no longer consistent with either the American CE or PE systems.

The uniqueness of the Japanese Gijutsushi system faced challenges of inconsistency with the international standard of CE. When the JCEA sought to join FIDIC circa 1960, they were rejected for their lack of independence and neutrality. Its affiliation with FIDIC was eventually achieved by establishing the AJCE, another Japanese CE association, in 1974. However, the international inconsistency of the Gijutsushi system remained as a "CE/PE controversy" because the system was not reformed in the 1970s (IPEJ 2001b, 43).

In the 1990s, another instance of international equivalence emerged from beyond the JCEA. At the conference of the World Federation of Engineering Organizations (WFEO), Matsumoto Junichiro, Japan's WFEO representative, was criticized because "the PE license, which is internationally recognized as an engineers' qualification, does not exist in Japan." Subsequently, at the June 1991 JFES board meeting, he submitted an agenda that declared it necessary to establish a new engineering license system in Japan (Kimura 1992; Saeki 2007).[1] Moreover, some companies expanding overseas had experienced similar troubles since the 1980s.[2] In particular, Imai Kaneichiro at IHI Corporation had witnessed young Japanese engineers without a PE license refused participation in the international project to develop the jet engine V2500 launched by the United States, the United Kingdom, and Japan in 1982 (Imai 2011; Imai, interview by the author, June 23, 2012; Hirose 2011).[3] Moreover, the Washington Accord was signed in 1989. Thus, to cope with the global standardization, the Japan Technology Transfer Association (JTTAS), where Imai served as the director, decided to introduce a new system circa 1992 to offer Japanese engineers opportunities to take the American PE examination and register for its license within Japan.

The JCEA was threatened by these campaigns and organized the External Affairs Committee to address the predicament in March 1992. The core issue of this committee was stated as follows:

> Although Japan became a major economic power, high-level engineers are far from fully recognized, despite serving as the real promoters. Improving the authority and the status of Gijutsushi as representative engineers has long been sought but never realized. (JCEA External Affairs Committee 1994a)

Given the complaint regarding their social status, the committee's policies were established to (1) increase the Gijutsushi's authority, (2) improve the social recognition of Gijutsushi, (3) expand Gijutsushi's business, and (4) ensure involvement in international businesses (JCEA 1992a). Moreover, the committee was divided into five subcommittees to address issues, including those related to social status. The PE issue was assigned to the First Subcommittee,[4] setting the objective "to counter the campaign to establish a new national license of Professional Engineer (if this license had been established separately from Gijutsushi, Gijutsushi would be denied the designation of PE)" (Ogawa 1992). In November 1992, the First Subcommittee published an urgent recommendation: "The JCEA should be the main agency in dealing with the international consistency of engineers' qualifications by amending the law to certify that Gijutsushi is not just PE but also CE if registered in the office with neutrality" (Kimura 1992; JCEA 1992b).

However, in January 1993, the JTTAS announced a plan to set up the Japan PE Council (Nihon PE Kyōgikai), a different organization from the Japan PE/FE Examiners Council, to provide examination and registration services for the American PE license within Japan in association with the National Council of Examiners for Engineering and Surveying (NCEES) of the United States. This plan was reported in a newspaper on January 19, 1993, as "introducing a professional license system for engineers: the JTTAS established a council with industry-government-academia cooperation to carry out a plan entrusted by the United States" (*Japan Chemical Daily* 1993). It urged the JCEA External Affairs Committee to negotiate with the JTTAS and NCEES (JCEA External Affairs Committee 1993, 1994c). The PE plan was not realized due to the objections at the NCEES General Meeting. However, at the same meeting, Imai persuaded the state of Oregon to hold their Fundamentals of Engineering (FE) examination in Tokyo (Imai 2011; Japan PE Council 1994, 20–26).[5] *Nikkei Business Daily* reported this examination in October 1994 as "Paving a Way for the Objective Qualifications for Engineers" and criticized the Gijutsushi system as follows:

> Although Japan has the Gijutsushi qualification, which is similar to PE, it is like a title provided to veteran engineers who "achieved distinction and fame." An official of the Japan PE Council asserted that "there is no measure in Japan to judge the ability of engineers." (*Nikkei Business Daily* 1994)

The JTTAS planned to offer the convenience of taking the American PE examination and registering its license in Japan without traveling to the United States.[6] Although this PE license was an American system, it became a global standard via the Washington Accord. The JCEA recognized the JTTAS' initiative as "a plan to seriously disrupt the sound development and operation of the Gijutsushi system" because a newspaper published such criticism when the JCEA was concerned with the possibility of establishing another engineering license in Japan due to the international inconsistency of Gijutsushi (JCEA External Affairs Committee 1993, 21, 1994b).

The JCEA's fiftieth-anniversary book reported that WFEO "asked about the linkage between engineering education and engineers' qualifications in Japan, the SCJ and JFES began to examine this problem, and the JCEA was worried about them affecting the Gijutsushi system" (IPEJ 2001b, 189). However, while the JFES board meeting discussed "the Issue of WFEO" as an agenda item in June and July 1991, such a plan made little progress (JFES 1992, 8–9; Ohashi Hideo, interview by the author, August 4, 2014). In contrast, an attempt to identify the Gijutsushi as a PE license was met with strong opposition within the JCEA because they believed that the Gijutsushi's qualification level was higher than that of PE (Kimura 1992).[7] Therefore, it was unavoidable for the JCEA to resolve its preexisting CE/PE controversy before resolving the new international problem (Kimura 2007). However, in 1994, the JCEA reached a consensus that "the Gijutsushi is an internationally comparable qualification of engineers, and respecting each country's qualifications, international mutual recognition will be inevitable in the near future" (JCEA External Affairs Committee 1994b). Moreover, it recognized the consensus as a problem "in the near future," but not as a current issue. Thus, they suggested changing the English name and promoting themselves to the industry, government, and academia rather than pursuing a radical overhaul of the system (Hori 1994b, 17).[8]

Taki Shigeatsu, a Gijutsushi, actively worked as a vice-chairperson of the First Subcommittee and repeatedly published articles on the PE issues in the JCEA journal.[9] In an article in 1993, he concluded that the Gijutsushi almost had international equivalence, except in the graduation requirement for an engineering college, relative to the American PE and the British Chartered Engineer. He supposed that the deficiency had reduced the SCJ's estimation of the Gijutsushi and suggested that:

> It will be necessary to examine the curricula of engineering colleges in associations with the SCJ and others and to reconstruct them to a double-track examination system, which became a requirement not only for graduation but also for taking the Gijutsushi's qualification examination. (Taki 1993)

Taki's idea in 1993 was quite similar to the establishment of JABEE and the amendment of the Japanese CE Act in the late 1990s.

After leaving JGC Corporation, Taki stayed at the East-West Center in Hawaii from September 1993 to February 1994 as a visiting researcher and increased his knowledge of the American PE system. In his report on PE, he mentioned the relationship between the rights of engineers and professional ethics: "In order to acquire their rights as a profession, American PEs organized a society to eliminate unqualified competition, prepare for professional ethics, improve their engineering skills, and ensure a proper reward" (Taki

1994). Furthermore, he praised their deep will to contribute to society and self-improvement. After attending a government-related conference of the American Association of Engineering Societies in March 1995, he reported on "The Role of Engineers in Sustainable Development" and explained the importance of a code of conduct and ethics from that perspective (Taki 1995).

Later, in the late 1990s, Taki's attention to professional ethics led to the introduction of engineering ethics in the JCEA. However, the JCEA's interest in engineering ethics remained limited in the mid-1990s. For example, the chairperson of the JCEA External Affairs Committee emphasized the philosophy of technology in the "Gijutsushi's Identity," but he treated ethics only as a part of Plato's natural philosophy (Hori 1994a).

GIJUTSUSHI'S SOCIAL STATUS AND ETHICS IN APEC MULTILATERALISM

When the JCEA was shaken by the international inconsistency issue, Saito Tetsuo, a Gijutsushi and the House of Representatives member first elected in 1993, repeatedly complained about the lower social recognition of Gijutsushi. The impetus was not only the inconsistency issue but also the fact that the Japanese CE Act was amended in 1983 to introduce the Associate Gijutsushi system to young engineers and delegate the examination and registration services to a private organization (i.e., JCEA) but without improvement in the use of the system and its social recognition (IPEJ 2001b, 18–22). Generally, the *White Paper on Science and Technology 1989* noted how science and engineering students avoid the manufacturing industry. More fundamentally, the *White Paper 1993* reported that young students avoid the study of science and technology (*wakamono no kagaku gijutsu banare*); thus, it became a major social issue, of which low social status and inadequate treatment of engineers were suspected to be among the causes.

At the February 1995 meeting of the Committee on Science and Technology of the House of Representatives, Saito appealed to improve the poor recognition of the Japanese Gijutsushi system by presenting the results of a survey of 5,000 engineers working at companies conducted by the JCEA and commissioned by the STA. The results demonstrated that low salaries were the most significant issue, followed by engineers' social status (Committee on Science and Technology 2 (132nd Diet Rec. Feb. 15, 1995)).[10] He also proposed the realization of mutual recognition between the Gijutsushi system and the American PE, arguing that the Japanese system was not currently accepted worldwide. In response to Saito's proposal, the STA secretary and the director-general of the STA Science and Technology Promotion Bureau responded that the Gijutsushi system was not restricted to consulting services but aimed to prove engineers' advanced abilities in science and technology. Moreover, it vaguely promised to advertise the Gijutsushi system and work toward realizing an international mutual recognition (Committee on Science and Technology 2 (132nd Diet Rec. Feb. 15, 1995)).

When Saito made a similar claim to the committee in March 1998, in contrast to three years prior, the STA secretary responded with reference to the active promotion of the APEC Engineer Register that there was a movement to realize engineers' mobility through international mutual recognition (Committee on Science and Technology 2 (142nd Diet Rec. Mar. 11, 1998)). Although the inconsistency issue with the American PE did not help reform the Gijutsushi system, the APEC Engineer issue, another external factor, had to be addressed at a national level, including the system's reform.

The Asia-Pacific Economic Cooperation (APEC) was established in November 1989. Amid the collapse of the Cold War's politico-economic structure, the socialist states were expected to shift to a market economy and be integrated into a global market. Conversely, the General Agreement on Tariffs and Trade (GATT) Uruguay Round, which began in 1986, faced difficult negotiations, raising concerns about economic globalization. In response, a free trade agreement (FTA) for each region was formulated. In North America, for example, the Canada-U.S. Free Trade Agreement came into force in 1989, and the North American Free Trade Agreement (NAFTA) in 1994, with the addition of Mexico. In Europe, an agreement for the European Union was established in 1993, and one for the European Economic Area came into force in 1994 along with the existing European Free Trade Association. In Southeast Asia, the ASEAN Free Trade Area agreement came into force in 1993.

The expansion of the *tight* regionalism caused concerns regarding the expansion of enclosing economic blocs. Moreover, the expansion of protectionist unilateralism raised concerns of retaliatory measures, represented by Section 301 of the U.S. Trade Act (called "Super 301"). In Japan-U.S. relations, resolving structural problems became a major diplomatic issue. Hence, APEC enacted a unique philosophy of "open regionalism" to ensure consistency with the GATT's multilateral system and complement and boost economic globalization.[11] Although the "open" and "regionalism" paradigms seem to be at odds, they were expected to function side by side to prevent exclusive bloc economies by providing a most-favored-nation treatment to external countries (Yamagami 1994, 112–116).

Subsequently, the GATT Uruguay Round reached an agreement in 1994; the World Trade Organization (WTO) was established in 1995. It made the trade in services and intellectual property rights a subject of trade negotiations, integrating engineering services into a global-free trade market. International mutual recognition of *substantial equivalence* of engineering qualifications and education had been arranged in multilateral agreements since the late 1980s, as discussed in chapter 5. APEC also promoted technology cooperation and transfer inside the region since its inception. In the Osaka Action Agenda in November 1995, APEC declared the promotion of business people's mobility and human resource development and confirmed that members should "seek to expand and develop researcher exchange schemes and engineer training schemes" in industrial science and technology (APEC 1995b). At the Human Resource Development Working Group (HRDWG) in January 1996, in response to the agenda, the mutual recognition project for the APEC Engineer Register was proposed by Australia, the host country, and was approved.

The fact that the APEC Engineer Register was an agreement on the PE licenses, not for CE services, was a serious issue for Japan. Following the APEC's approval, the STA's Liaison Committee on Engineers' Licensing Issue (Gijutsusha Shikaku Mondai Renraku Kondankai) was established in May 1996. The Gijutsushi system, under the jurisdiction of the STA, was coordinated as a corresponding license to APEC Engineer, but various problems remained (Ohashi, interview by the author, August 4, 2014). The JFES had been aware of the Gijutsushi's international inconsistency since the early 1990s, as described in the previous section, and considered creating another new license titled "engineer" at the Special Committee on Increasing Engineers' Esteem (Gijutsusha no Shakaiteki Hyōka Kōjō ni tsuite no Tokubetsu Iinkai) established in February 1995 (JFES 1995).[12]

Under these circumstances, the JCEA surveyed relevant ministries and academic societies from December 1996 to February 1997 and delivered a statement that the representative Japanese licenses for engineers were limited to Gijutsushi and Kenchikushi (Registered Architect) (IPEJ 2001b, 190).[13] Additionally, because the JCEA External Affairs Committee was a temporary committee not stipulated in the JCEA's articles and bylaws, the JCEA board established a new specialized committee on the qualifications issue in September 1996 and transferred other duties to other standing committees. The board then organized the Research Committee on Engineers' Licensing Issue (Gijutsusha Shikaku Mondai Chōsa Iinkai) in November 1997 (IPEJ 2001b, 91, 103–104).

Concurrently, the APEC's impact accelerated the survey of accreditation of engineering education, which had been conducted since the early 1990s. In November 1994, the WFEO Committee on Education and Training (1994, 3–12) decided to recommend that member countries implement engineering education accreditation; it submitted its recommendation in 1995.[14] Although the generalization of the Standards for Establishment of Universities (SEU) introduced self-evaluation in college education in Japan in 1991, WFEO required them to introduce an accreditation system with mutual evaluation by an independent organization per the global standard (Imai 1994, 1995).[15] In response to the requests, the JFES established the Study Committee on the Need for Evaluation of University Engineering Education Programs (Daigaku Kōgaku Kyōiku Program Hyōka no Hitsuyōsei Kentō Iinkai) in November 1995. The JSEE also established the Research Committee on the Accreditation System for Engineering Education (Kōgaku Kyōiku Accreditation System Chōsa Kenkyū Iinkai) in July 1996. After a year of coordination, the JSEE and the JFES established the Arrangement Committee for Engineering Education with International Consistency (Kokusaiteki ni Tsūyōsuru Engineer Kyōiku Kentō Iinkai) in July 1997. This committee, attended by members of the industry, academia, and government, including the JCEA, organized the establishment of JABEE. The details of the process are discussed in the next chapter.

Preparations for the APEC Engineer Register also proceeded steadily. At the HRDWG workshop in Manila in August 1997, the qualification criteria for APEC Engineers were agreed upon as follows:

completed an accredited or recognised engineering program, or assessed recognised equivalent; and been assessed within their own economy as eligible for independent practice; and gained a minimum of seven years practical experience since graduation; and spent at least two years in responsible charge of significant engineering work; and maintained their continuing professional development at a satisfactory level. (APEC 2000, 7)

These criteria were initially discussed on the condition of engineers being *accredited* by an independent organization; however, there was a possibility that none of the existing Japanese engineers could meet the criteria. Therefore, Japan negotiated to include those *recognized* by government agencies, such as the MOE, and the phrase "or recognised" was added to "accredited" (*Kensetsu Tsushin Shimbun* 1997).[16] Taki of the JCEA chaired the subcommittee at the Manila workshop. In the JCEA, Taki became responsible for the APEC Engineer project in May 1996, became a board member in June 1997, and was actively involved in the APEC Engineer issue.[17]

In addition to these five criteria, two additional requirements in the code of ethics and accountability were defined: "bound by the codes of professional conduct established and enforced by their home jurisdiction and by any other jurisdiction within which they practice" and "held individually accountable for their actions, both through requirements imposed by the licensing or registering body in the jurisdictions in which they work and through legal processes" (APEC 2000, 7–8). Compliance with the codes of conduct in the involved countries was also emphasized in the guideline (APEC 2000, 14).[18] The fact that engineering ethics became a qualification requirement for the

APEC Engineer caused Japanese society to recognize its necessity. In June 1997, prior to the APEC Engineer agreement, the WFEO Subcommittee of the Liaison Committee on Basic Engineering Research (Kiso Kōgaku Kenkyū Renraku Iinkai) in the SCJ's Section 5 (on engineering) published the *Proposal for Ethics Education for Engineers at Higher Education Institutions in Engineering* (*Kōgakukei kōtō kyōiku kikan deno gijutsusha no rinri kyōiku ni kansuru teian*) and distributed it to relevant ministries and engineering faculties. However, they received only a single response to it (Nishino 2004). The proposal recommended not only the establishment of a new code of ethics per the globalization and development of science and technology but also the introduction of engineering ethics education in college education modeled on ABET (SCJ 1997).

Nishino Fumio, Japan's representative for WFEO and the chairperson of the WFEO Subcommittee of the SCJ's Section 5, led the drafting of the SCJ's proposal.[19] Following the adoption of the WFEO Code of Environmental Ethics for Engineers in 1987, WFEO began developing a model code of ethics in 1990. Matsumoto Junichiro, Nishino's predecessor as Japan's representative for WFEO, claimed the following regarding the code of ethics at the WFEO meeting: "Only Japan, China, and Korea did not have such a thing among the big countries" (EAJ 1996, 9).[20] However, rather than the WFEO's initiative, he advocated for the importance of engineering ethics via the subcommittee, which he chaired at the SCJ (Ohashi, interview by the author, August 4, 2014). His belief is reflected in the nature of the proposal. Nishino had a strong interest in internationalizing engineering licenses and education, especially in the East and Southeast Asia region. Moreover, he chaired the Study Committee on Mutual Recognition of Engineering License (Gijutsusha Shikaku no Sōgo Shōnintō Kentōkai), established by the Ministry of Construction in December 1996 to address the APEC Engineer issue.

Fudano Jun at Kanazawa Institute of Technology (KIT) introduced the situation of American engineering ethics education to the WFEO Subcommittee.[21] Meanwhile, a nationwide reorganization of liberal arts education progressed under the influence of the SEU's generalization in 1991, and KIT had been implementing education reforms since 1995, modeled after American engineering education. He had focused on American engineering ethics education as a node of both perspectives. KIT's initiative was a unique case in Japan. As of June 1997, when the proposal was issued, the social awareness of engineering ethics education was quite low. KIT's efforts are further discussed in chapter 7.

The JSEE and JFES Arrangement Committee for Engineering Education was established in July 1997. Umeda Masao, president of the JCEA, recalls that the SCJ, the JFES, the MOE, the STA, the MITI, and Keidanren set up the Committee to Produce Engineers with International Consistency (Sekai ni

Tsūyōsuru Gijutsusha o Tsukuru Iinkai). Moreover, they launched a campaign designed to create a new engineering license to join the Engineers Mobility Forum (EMF), which was introduced in October 1997 as an international agreement under the Washington Accord, in addition to the APEC Engineer agreement (Umeda 2007, 10–11). This situation threatened the social status of Gijutsushi in the extension of the problem of the early 1990s. However, as the title suggests, the committee was solely for engineering education. The first steering meeting agreed that "the issue of professional engineer license is outside the scope of this committee and will not be discussed" (JABEE 2003, 13–14). Ohashi Hideo, the vice-chairperson, also stated that the committee did not aim to establish a new engineering license and was not considering the EMF (Ohashi, interview by the author, August 4, 2014). However, the possibility of establishing a license management organization for the engineers' registration service was mentioned at the third steering meeting.[22] In a bitter recollection of the early 1990s, the JCEA could not dispel concerns that the effort to establish an accreditation body of engineering education with international consistency would also lead to the establishment of a new engineering license.

Amid this urgent situation, the APEC Engineer scheme was confirmed at the HRDWG in June 1998, just before finalizing the APEC Engineer license in each country in November. In September 1998, JCEA President Umeda, the secretary-general, and Taki visited Ishikawa Rokuro, president of the JFES, and Yoshikawa Hiroyuki, president of the SCJ and the chairperson of the JSEE and JFES joint committee, and successfully petitioned for the Gijutsushi's placement in the APEC Engineer agreement. While Yoshikawa was troubled by the APEC's engineering ethics requirement, they prepared the Japanese translation of *Engineering Ethics: Concepts and Cases*, just a week before its publication on September 30 and brought it to the meeting. Thus, the Gijutsushi was finally recognized as appropriate for the APEC Engineer license for fulfilling the requirements (Umeda 2007, 11–19).

After the November 1998 steering meeting, the APEC/HRDWG entered the establishment stage. The JCEA then revised the existing Code of Ethics for CE Practice (Gijutsushi Gyōmu Rinri Yōkō) to that for PEs (Gijutsushi Rinri Yōkō) in March 1999. The preamble was extensively revised as follows:

> As professional engineers, we hold paramount the safety, health and welfare of the public, recognize our professional mission, social status, and obligations, continuously improve our competence, maintain neutrality and impartiality, have pride as selected professional engineers, and endeavor to perform this code of ethics. (IPEJ 2001b, 283)

Thus, public interest requirements for PEs were clearly specified, and "neutrality of the position" regarding CE was changed to "maintenance of neutrality and impartiality" in consideration of PE. In February 2000, the STA's Gijutsushi Council submitted a proposal, *On Measures to Improve the Gijutsushi System*, based on the STA's Liaison Committee's report, *The Necessity of Responding to the Mutual Recognition Project of APEC Engineer and the Improvement of the Gijutsushi System*, in June 1999.[23] This proposal listed professional ethics at the top of the measures and requested to "clarify and ensure social responsibility as the professional engineer's prerequisite to conducting business so as not to prevent public interest, such as public safety and environmental conservation" and "ensure professional ethics, such as obtaining appropriate professional assistance when exceeding the ability, through examination and continuing education" (Gijutsushi Council 2000).

JABEE was then established in November 1999, and the Japanese CE Act was amended to the PE Act in April 2000. In this amendment, two articles were added as responsibilities for public interests and development in professional competence, as with the code of ethics. Article 45 (2): "No PE nor Associate PE shall engage in operations that harm public interests, including public safety, environmental preservation, etc., during normal business operations." Article 47 (2): "PEs shall, at all times, endeavor to continually increase their knowledge and develop their skills with regard to their profession." Moreover, the equivalence between Gijutsushi and the other engineering licenses of other countries permitted by the MOE minister and that between the Associate PE and the graduates of the educational institutions designated by the MOE minister (actually the engineering colleges accredited by JABEE) were added as special cases (*Kanpo Official Gazette* 2000, 6–7). When the APEC Engineer registration service began in November 2000, the English titles of Gijutsushi and the Gijutsushi-kai changed from C.E. and the JCEA to P.E.Jp and the Institution of Professional Engineers, Japan (IPEJ), respectively.

The concept of international mutual recognition for engineering licenses is a matter of the WTO General Agreement on Trade in Services (GATS). The APEC Engineer Register followed the WTO's idea. In the late 1990s, however, engineer licensing was managed not as a WTO issue but as an APEC Engineer issue. When the WTO Agreement on Government Procurement came into force in January 1996, the Ministries of Construction; Transport; and Agriculture, Forestry, and Fisheries revised the standard terms and conditions of contracts on the management and verification engineers as Gijutsushi; no other WTO coordination was required. Nishino explained that WTO/GATS did not promote the mutual recognition of engineering licenses because establishing the agreement was difficult due to the difference in qualification requirements in each country. Furthermore, governments had dire concerns other than engineers, such as accountants, lawyers, and architects

(Nishino 1997a, 1997b). Therefore, the WTO did not present any specific discussions on engineering ethics.

TRANSLATION OF THE ENGINEERING ETHICS TEXTBOOK

In 1998, the JCEA/TTC translated and published the textbook, *Engineering Ethics: Concepts and Cases*. This group was organized in January 1993 under the name "Gijutsushi Hon'yaku Center" (Japanese CE's Translation Center), a JCEA's project team.[24] It aimed to translate overseas technical documents in anticipation of the growing demands from Japan and reduce barriers to promoting technology access from other countries and technology transfer from Japan. Kudo Hisha,[25] the TTC's project leader, described the purpose as follows:

> In recent years, Japanese engineers have been able to refer to excellent English technical documents and patents besides Japanese ones; however, Western engineers have not been able to sufficiently refer to technical documents written in Japanese. One of the reasons that Japan has become the world's best country, especially in civil [not military] technologies, seems to be attributed to the difference in the engineers' language skills. With the advancement of Japan's industrial technologies, overseas expansions have become popular, and engineers with language skills are required to conduct business. Foreign languages are indispensable for introducing the technologies developed in Japan to the world. (Kudo 1995, 17)

Accordingly, the TTC's activities included collaboration among CEs engaged in translation services, planning and development of translation services, support for dispute resolution, and research on translation skills by introducing commissioned work and work experience (Kudo 1995). As the team's purpose regarded translation, the topic of translation was not restricted to specific areas, such as engineering ethics. Until 1996, for example, they dealt with a variety of topics, such as American PE, food for specified health uses (FOSHU), the Product Liability Act, and ISO 9000 and 14000, which became issues at the JCEA.[26] Engineering ethics was not initially envisioned within the scope of translation.

In the fall of 1995, while topics such as ISO 9000 and new product development strategy were considered as candidates for translation, "in his briefing on ABET, APEC, NSPE, and so on," Taki suggested a translation of engineering ethics because "engineering ethics will become indispensable for the faculty of engineering" (Fukumoto Muneki, interview by the author, July 13,

2012; email message to author, July 26, 2012).[27] However, the proposal was not approved by the TTC in 1995. Apart from Taki, another member likely proposed to translate Martin and Schinzinger's textbook (1996) to the JCEA secretariat, but the plan did not evolve further due to the challenging content (Hatakeyama Masaki, interview by the author, August 31, 2012; email message to author, September 1, 2012).

At the monthly meeting in September 1997, the plan to translate the engineering ethics textbook began to take shape. Morihiro Eiichi (2008, 18) proposed "On *Ethics* translation," recorded in the minutes as the English word "Ethics," as a common issue in the Gijutsushi's technical disciplines (Hattori 1997b). The Ethics Research Subcommittee (Ethics Kenkyū Bunkakai) for the translation was then organized with eight people: Sugimoto Taiji as the leader, Fukumoto Muneki, Kashima Minoru, Kogami Kunio, Kobayashi Hiroomi, Kudo, Morihiro, and Sakurai Koichi (JCEA 2002, 423–424).[28] Taki was also an adviser to the group (JCEA 1998). At the APEC/HRDWG workshop held the previous month in August 1997, as mentioned in the previous section, a basic agreement was reached to make ethics a requirement for the APEC Engineer license, and Taki was the person in charge of negotiating the APEC Engineer scheme at the workshop and other meetings.[29]

Taki suggested approximately ten books as candidates for translation, and members circulated these for selection. *Engineering Law, Design Liability, and Professional Ethics* by Rebecca J. Morton (1983), which had only three pages on ethics out of eighty-seven pages,[30] emerged as a candidate but did not obtain member approval. Thus, at Taki's suggestion, they decided to seek advice from Arthur Schwartz, an NSPE specialist in engineering ethics and legal affairs, at the November 1997 meeting. Morihiro first selected *Engineering Ethics: Concepts and Cases* (Fukumoto, Hamada Tetsuo, and Kurosawa Takeo, interview by the author, July 13, 2012). They then confirmed the selection's validity with Fudano of KIT in addition to Schwartz's recommendation. At the December meeting, the TTC decided to translate the book following Kudo's opinion: "case-centric, interesting to read, practical, and motivating to purchase" (JCEA/TTC 1998; Fukumoto, interview by the author, July 13, 2012). The members then shared the book's translation and published it on September 30, 1998, under the Japanese title "Kagaku gijutsusha no rinri" (Ethics for Science-Engineers).[31]

Machine translation was actively employed in translating the book. First, the TTC's purpose was similar to that of machine translation development in Japan in the 1980s. Japan's machine translation began with STA's Mu Project, conducted for four years beginning in 1982.[32] The goal was to share information on the abstracts of scientific papers between the United States and Japan to curb the U.S.-Japan trade friction (Nagao 2010, 92–93). From 1986, the project evolved into an interdisciplinary research project, involving engineering, linguistics, and psychology researchers, as a KAKENHI Special Project Research. In 1987, the Machine Translation System Laboratory was also established at the MITI Center of the International Cooperation for Computerization. At this time, the Multilingual Machine Translation Project was conducted to promote technology transfer as part of Japan's Official Development Assistance (ODA) in five countries: Japan, China, Indonesia, Malaysia, and Thailand. Although the JCEA was uninvolved in the projects, members' overseas businesses regarded the technology transfer through Japan's ODA.[33]

Electricity and telecommunications companies that participated in the national projects actively promoted the commercialization of machine translation technologies. In the late 1980s, the translation system, bundled with hardware, cost several million yen; the software alone cost approximately one million yen. Since circa 1990, however, prices had dropped drastically due to technological advances. Specifically, the translation software, Korya Eiwa,[34] sold by LogoVista in 1994, was priced less than 10,000 yen; since then, machine translation has been popular not only with companies but also with individuals (Narita 1997, 112–127).

Per technological advances, the TTC first took up the topic "machine translation using OCR [optical character reader]" at its monthly meeting in October 1994 (Kudo 1995, 18). In the September 1995 article, Kudo recommended that preparing office automation equipment, such as desktop publishing (DTP) and PC communication, improved the efficiency of editing print layouts and sending documents. Furthermore, combining OCR and machine translation improved the efficiency of translating and learning a second language (Kudo 1995, 19–20). Thus, considering the purpose of technology transfer, translations between English and other Western European or Russian languages were also expected. Given the strong interest in machine translation, Kudo made efforts to popularize it while advising TTC members on preparing a PC and Internet environment (Fukumoto, Hamada, and Kurosawa, interview by the author, July 13, 2012). For example, in late August 1997, the TTC held a two-day training camp on the themes, "presentation of overseas experience in English for thirty minutes" and "hands-on experience of machine translation" at Kudo's office in Tsukuba (Hattori 1997a).

The JCEA's fiftieth-anniversary book described the use of the computer system as follows:

> Thus, with the large 400 pages of the original, the 460 pages of the translation became a book at an astonishing speed, as the translation started in January 1998, and the book was published in September. This speed was attributed to members' contribution with computer aided work, the use of Hitachi's English-to-Japanese translation system Tachimachi Hon'yaku [Immediate Translation],

the use of the DTP software from translating to typesetting, and the cooperation of the Publishing Division of Maruzen Co., Ltd. (IPEJ 2001b, 221).

The acknowledgment of the book clearly stated that Hitachi's translation software and Adobe's DTP software, PageMaker, were used for translation and editing (JCEA 1998). The price of Hitachi's software was also less than 10,000 yen at the time.[35] Their April 1998 progress report proudly described the use of machine translation: "In this subcommittee, we have divided the translation work; many people have completed their work by using machine translation efficiently" (Hattori 1998). Machine translation was not compulsory; however, its definite acknowledgment suggests that it was vital in their activities and was not merely a convenient auxiliary means.

As described above, the translation and publication of *Engineering Ethics: Concepts and Cases* were conducted by the TTC volunteers who initially did not focus on engineering ethics. The theme of engineering ethics education was external and incidental. An introductory article on the TTC's activities in 2005 stated that this book was the first Japanese textbook on engineering ethics; the subject was also emphasized in their subsequent activities because, for example, the book was praised as a textbook for ethics lectures at engineering colleges (Kobayashi 2005, 29). Engineering ethics was initially only an option, and the subsequent translation publications on the subject were largely due to the success of the publishing business. Engineering ethics education garnered little attention until the middle of 1997, and its importance has been recognized since the basic agreement was reached in August 1997 to make ethics a qualification requirement for the APEC Engineer agreement. This occasion was no exception in the TTC. A license agreement for translation publishing was signed between the JCEA and Wadsworth Publishing Company between late 1997 and January 1998. Soon after, however, another Japanese publisher asked Wadsworth to license the translation of the book (JCEA/TTC 1998; Fukumoto, interview by the author, July 13, 2012). Hence, the JCEA was not much farther along in the situation.

ENGINEERING ETHICS AS A GLOBAL STANDARD FOR ENGINEERS' SOCIAL STATUS

Regarding the early JCEA, which established the codes of ethics in 1951 and 1961, international inconsistency was not an urgent issue to be solved. Engineering ethics education had neither been introduced nor recognized as a requirement. In the 1980s, compliance became strongly demanded, given corporate misconducts such as the Toshiba Machine–Kongsberg violation of CoCom in 1987, the amendment of the Securities and Exchange Act in 1989, and the U.S.-Japan Structural Impediments Initiative (SII) talks. However, professional ethics education for engineers had not become a social issue in Japan. The U.S.-Japan Joint Task Force, which JSPS Committee 149 engaged in, discussed neither the reform of the Japanese Gijutsushi system nor the requirement for engineering ethics, as described in chapter 5.

The JCEA's translation and publication of engineering ethics education material were closely related to the various efforts of coordination with the APEC Engineer project inside and outside the JCEA. Behind this situation was the national challenge of responding to the new economic framework in the Asia-Pacific region. During this period, the world economy's philosophy was at a major turning point from bilateralism, such as in postwar U.S.-Japan relations, to multilateralism, such as in APEC. In particular, maintaining economic initiatives in the Asia-Pacific region became an important issue for Japan, which had become a developed country. Therefore, the international equivalence issue constituted external pressure to the JCEA from inside Japan rather than outside. Even if the pressure was not addressed, it was strongly suspected that not only the overseas but also the domestic social status of Gijutsushi would collapse. Thus, to secure social status in compliance with the new global framework, the JCEA focused on engineering ethics education and revised the code of ethics.

The JCEA's fiftieth-anniversary book praised its translation and publication of the textbook as "a success of the socially outward-looking activity by the often inward-looking JCEA" (IPEJ 2001b, 221). The introversive JCEA's recognizing the need for and succeeding through socially extroversive work was derived from the TTC's focus on international demands and their reception of direct advice from Taki, who led the discussions on the APEC Engineer project in the turbulent situation.

The JCEA had been waking to a sense of crisis over the international inconsistency issue between Gijutsushi and the American PE since the early 1990s. Taki paid attention to the importance of engineering ethics education as the committee's vice-chair to address the issue at the early stage. The international inconsistency was treated as a matter of mutual understanding while it was a bilateral issue with the United States. However, institutional reform was required when it became a multilateral issue of the APEC Engineers' mutual recognition. Furthermore, the introduction of engineering ethics to Japan was inevitable when it became a requirement. Until the early 1990s, the main issue of internationalization in Japan was access to, cooperation in, and transfer of Japanese technologies. The JCEA/TTC was also launched based on these interests. However, in the late 1990s, the issue was to accommodate the global standards growing outside Japan; the direction required in the internationalization was reversed. Thus, the TTC pursued the "voluntary" and "socially outward-looking" initiative in a new extrinsic field of engineering ethics in translating and publishing the American textbook, utilizing machine translation.

Although the JCEA focused on the global scheme of the APEC Engineer agreement, the TTC translated ethics contents from American engineering education. It is natural to think that the ethics education on global engineering business should be published; however, they translated the American book because the APEC Engineer Register was premised on the accreditation agreement of the Washington Accord because they had recognized the United States as a leader in engineering qualifications through the international inconsistency issue. Moreover, there were few textbooks on engineering ethics outside the United States in the first place.

After all, even in the 1990s, as in the previous two codes of ethics, the direct motivation to introduce engineering ethics was to maintain and improve their social status. The JCEA's elitist and conservative tendencies, as discussed in chapter 3, also affected this process. Initially, only the establishment of the codes of ethics, not the education, was introduced. However, a more substantial implementation was demanded in the late 1990s. Thus, not only the international equivalence between engineering licenses but also the reform of the engineering education system per the global standard became an important issue for Japan. Hence, implementing engineering ethics education became an essential requirement. The next chapter discusses it in more detail.

NOTES

1. The JCEA's fiftieth-anniversary book reported that the JCEA obtained information on the JFES through Saeki, a JFES board member (IPEJ 2001b, 260), but the details of the proposal are uncertain.

2. The Engineering Advancement Association of Japan (ENAA) of the MITI also organized a committee consisting of various companies and conducted surveys on engineering qualifications in twenty-five countries worldwide from 1988 to 1990 (JMF and ENAA 1991).

3. This project was announced among the three countries in 1982 and entered into joint venture agreement in 1984 among five countries, including Germany and Italy.

4. Meanwhile, the Registered Civil Consulting Engineering Manager (RCCM) license was established apart from Gijutsushi. Thus, the Second Subcommittee was in charge of addressing the issue. The Third Subcommittee addressed international cooperation through entities such as the Japan International Cooperation Agency, the United Nations, and Japan's Official Development Assistance (ODA). The Fourth Subcommittee addressed Gijutsushi's legal affairs. The Fifth Committee addressed social recognition and media reports.

5. Since 1996, Oregon's PE examination had also been held at the U.S. Naval Base in Yokosuka, Japan.

6. Thus, Sugimoto Taiji (1995, 13), a Gijutsushi, also commented: "Japan's opposition is no longer valid because it is like an American university conducting remote examination in Japan [. . .] after all, it is not a failure of the JTTAS people to be convinced of the importance of the PE license in relation with the United States," and the JCEA's "lack of understanding of the legal structure of the American PE Act probably caused this trouble."

7. Opinions persisted that Gijutsushi should be regarded as a CE based on the purpose of the establishment. Equating Gijutsushi with PE was thought to reduce the Gijutsushi's quality and social status. Even after the English title was specified as P.E.Jp in line with the APEC Engineer agreement, a considerable opposition remained (Takahashi 2007, 178). At the general meeting in 2001, a protest broke out against the English name being altered without sufficient consensus, and Taki persuaded the floor (Sato 2007).

8. Among the JFES, Uchida Moriya seemed skeptical of regarding Gijutsushi as a PE license, Kimura Katsusaburo, chairperson of the First Subcommittee, and Saeki Kazuyoshi and Taki, vice-chairperson, visited the JFES several times circa 1995 to persuade them (Kimura 2007; Saeki 2007).

9. Taki had long been interested in the American PE license. In 1991, he published the brief introduction in the JCEA journal and noted the possibility of moderating the U.S. claims for non-tariff barriers by achieving mutual recognition between PE and Gijutsushi (Taki 1991, 28). However, in the 1991 article, he only made a simple suggestion that he noticed in the course of daily business.

10. The report concluded that the fastest way to retain good engineers was to place the highest importance on improving payment and the working environment, which was considered to be "3K" (*kitsui, kitanai,* and *kiken*; hard, dirty, and dangerous). The *improvement of workplace environment and treatment* were represented in 73 out of 159 free answers, far exceeding 29 for the next highest-ranked *improvement of social status* (JCEA 1994, 113–114, 122–123, 126).

11. APEC set the objectives "to develop and strengthen the open multilateral trading system in the interest of Asia-Pacific and all other economies" and "to reduce barriers to trade in goods and services and investment among participants in a manner consistent with the GATT principles, where applicable, and without detriment to other economies" (APEC 1995a, 71).

12. This interim report reveals that the draft of the license was arranged.

13. In October 1997, the STA and the JCEA organized the Liaison Committee on the Gijutsushi Problem (Gijutsushi Mondai Renraku Kondankai). In March 1998, the JCEA submitted an interim report, *Basic Survey Report on Effective Utilization of Science and Technology Human Resources in the APEC Countries.*

14. This recommendation was made on April 18, 1995 (Carroll 1995; Harada 1996, 48–49).

15. Imai, who was a member of the WFEO Committee on Education and Training, noted that the engineers placed the highest priority on public safety, health, and

welfare, based on the American PE's engineering ethics, to fulfill their professional duties (Imai 1995, 3).

16. This article was based on an interview with the director of the Science, Technology, and Information Division of the STA. The JUAA, founded in 1947, did not function as an accreditation body required by the APEC Engineer project. In addition, it was also necessary to introduce CPD, which did not exist in the Japanese system.

17. Taki received the MEXT minister's award in 2003 for popularizing mutual recognition schemes to promote the international mobility of Japanese engineers.

18. The APEC's discussion paper in December 1997 specified the principal fields of approved education programs: mathematical and physical sciences, engineering sciences, and engineering analysis and design. It also specified the supplementary fields: communication, management, and ethics (APEC 2000, 30).

19. It was probably edited by Shibayama Tomoya, a member of the subcommittee. Both Nishino and Shibayama were former colleagues at the University of Tokyo's Faculty of Engineering.

20. This is a record of a study meeting held in Tokyo in July 1996. Matsumoto supposed that engineering ethics was not popular in Japan because collectivism, not individualism, was a feature of Japanese society.

21. He was introduced by Hori Yukio, a member of the subcommittee and the vice president of KIT, to participate in this proposal as a collaborator.

22. This idea was proposed as "the Japan Council for Professional Engineers," with accreditation, certification, and registration functions like in the United Kingdom and Canada, or as "the Japan Association for Professional Engineers," paired with the Japan Accreditation Board for Engineering Education" (JABEE 2003, 17–19). However, the possibility of such an organization did not appear except in the minutes of this third meeting.

23. The economic policy, approved by the Cabinet of Japan in July 1999, stimulated Japanese engineers by improving international consistency (Economic Planning Agency 1999).

24. The JCEA/TTC was organized mainly by old friends who played the game of Go and was established as the *Gijutsushi* Hon'yaku Center in January 1993 after the JCEA board approval in March 1992 (Kudo 1995; Fukumoto, Hamada, and Kurosawa, interview by the author, July 13, 2012). This name was changed to the *Gijutsu* Hon'yaku Center (Technical Translation Center, TTC) at the monthly meeting in March 1996 (Hattori 1996).

25. After working at Asahi Kasei Corporation, Kudo established the Tsukuba High-Tech Research Consultants in 1988, providing technical consulting services such as literature research, translation, and publishing.

26. See Kudo (1995, 18) and the reports of project teams in the JCEA journals.

27. Fukumoto's memorandum at the time did not record the specific date, but Taki's suggestion was probably made circa the fall of 1995 because the next schedule was set for November 25, 1995. As a continuation of the international consistency issue ongoing since the early 1990s, Taki was about to travel to Singapore, Malaysia,

Thailand, Australia, and the United States to investigate their engineer systems and introduce the Japanese system (IPEJ 2001b, 189).

28. Sugimoto, nominated as the leader, had a successful track record, having published a book on the PL Act, published with Kogami in 1996. This subcommittee was named the Engineering Ethics Subcommittee (Gijutsu Rinri Bunkakai).

29. Regarding the relationship between the APEC Engineer project and the translation of engineering ethics, another testimony notes that only a few JCEA members were aware of the international equivalence issue and that the importance of ethics was shared apart from the APEC issue (Hatakeyama, interview by the author, August 31, 2012). However, as Morihiro's proposal was written with the English word as *"Ethics* hon'yaku" (*Ethics* translation) in the minutes, and the TTC set up the translation group under the English name "Ethics Kenkyū Bunkakai" (the Ethics Research Subcommittee), it was evident that the supposed Japanese engineering ethics framework was not domestic but a Western idea. The JCEA's fiftieth-anniversary book stated that Morihiro "obtained the information on the United States where the engineering ethics education was emphasized, and proposed the translation of an appropriate textbook at the TTC monthly meeting around 1997" (IPEJ 2001b, 221). However, it is also unlikely that this was independent of the APEC context.

30. Only three pages in the book addressed the explanation of ethics, but seven pages of the appendix presented examples of typical codes of ethics.

31. Although the original title was *Engineering Ethics* without the word "science" (kagaku), the Japanese translation intentionally incorporated "science" to the title because, in addition to the understanding that technology is based on science, and engineering includes science, the aim was to expand the target readers from a marketing perspective (Fukumoto, Hamada, and Kurosawa, interview by the author, July 13, 2012).

32. This project was led by Nagao Makoto at the Faculty of Engineering, Kyoto University, along with the STA Japan Information Center of Science and Technology and the MITI Electrotechnical Laboratory and RIPS Center.

33. For the JCEA members' businesses through ODA, see IPEJ (2001b, 161–186, 194–212).

34. "Eiwa" means both *English-to-Japanese* and *good* in Japanese, therefore the title means "How a good English-to-Japanese translation!"

35. Furthermore, it was pre-installed on Hitachi's PCs.

Chapter 7

The Globalization of Engineering Education and Ethics

THE ESTABLISHMENT OF JABEE AND ENGINEERING ETHICS

As described in chapter 6, economic globalization in the 1990s required multilateral mutual recognition agreements for engineering qualifications with engineering ethics. In particular, the APEC Engineer agreement, demanding an accredited educational system, accelerated engineering education reform, which had become an issue by the early 1990s. Moreover, it motivated the establishment of the Japan Accreditation Board for Engineering Education (JABEE, Nihon Gijutsusha Kyōiku Nintei Kikō) in 1999. Since the mid-1990s, engineering ethics education has increasingly been emphasized as an essential requirement in the global standard. In addition to describing the introduction of ethics to engineering education in the late 1990s, this chapter discusses broad changes in the social recognition of American professional ethics during this period.

In Japan, educational accreditation has garnered attention since the late 1980s. In addition to the cases illustrated in chapter 5, Ohnaka Itsuo, an engineering professor at Osaka University, surveyed engineering education reform in the United States at the request of the MOE in 1994. He also invited ABET and NSF members, including ABET President Eleanor K. Baum, to an international symposium on the evaluation of engineering education in November 1995. Although WFEO recommended that member countries introduce an accreditation system in April 1995, the need for quality assurance of education was not well recognized in Japan. Furthermore, its introduction did not progress at the time (JABEE 2012, 21–22).[1] Moreover, the discussion focused on the institutional accreditation introduced by the JUAA in the postwar educational reform, which was not intended for professional training (Ohashi, interview by the author, May 12, 2012).

The introduction of accreditation linked to engineering qualification began with the approval of the Mutual Recognition Project of APEC Engineer in January 1996. In November 1995, the JFES established the Study Committee on the Need for Evaluation of University Engineering Education Programs. In July 1996, the JSEE went a step further and established the Research Committee on the Accreditation System for Engineering Education, chaired by Ohashi Hideo, president of Kogakuin University. At the beginning of the first JSEE Committee meeting, Ohashi explained that the committee had been established because of the requirement that APEC Engineers graduate from an accredited educational program. Moreover, he warned that failure to meet this requirement would disadvantage engineering graduates and severely harm Japan's national interests (Ohashi 1997, 51; JABEE 2003, 3). Although the SEU's generalization implemented the quality assessment of college education in 1991, they lacked the means to ensure compliance with APEC standards systematically. Ohashi also reported that "the motivation for the birth of this committee was due to the external factor of APEC" (Ohashi 1997, 52).[2] That is, the desire to maintain the social status of Japanese engineers in the global economy primarily drove the committee's establishment.

The JSEE Committee sought to implement an accreditation system in each engineering field to shift engineering education from conventional academic education to professional training. In June 1997, the JSEE Committee defined engineering education as "an education to develop human resources to engage in the engineering profession in the future, which includes not only specialized knowledge and skills but also human education to develop judgment, ethics, autonomy, and communication skills necessary for engineers" (JABEE 2003, 9). Eventually, JABEE's definition incorporated liberal arts education, design education, and practical training, which also required engineering ethics education (JABEE 2003, 31–32).

In the same year, the WFEO Subcommittee of the SCJ's Section 5 deliberated on the introduction of engineering ethics education, publishing a report in June 1997, as described in chapter 6. Both Ohashi and Ohnaka were board members. The report advanced three recommendations: (1) engineering ethics education should be provided in undergraduate education, (2) student abilities to make appropriate judgments should be cultivated through case studies rather than memorizing specific values, and (3) a code of ethics adapted to the needs of the times should be established in all engineering societies (SCJ 1997).[3] Using ABET as an example,[4] the report also maintained that it was necessary to introduce vocational ethics education for the engineering profession rather than merely appealing to a sense of mission.

Although the SCJ's report garnered little response at the time,[5] engineering societies began preparing codes of ethics, and the introduction of engineering ethics education gradually emerged as an issue in engineering education. When the SCJ entered its seventeenth term in July 1997, the JSEE and the JFES integrated their engineering education committees into the Arrangement Committee for Engineering Education with International Consistency (Kokusaiteki ni Tsūyōsuru Engineer Kyōiku Kentō Iinkai), chaired by the JSEE president, Yoshikawa Hiroyuki, who was also elected as the SCJ president in 1997. The joint committee advanced the more specific arrangements by establishing the Study Committee on Engineering Education (chaired by Ohashi), the Coordination Committee on Engineers' Licensing Issue (chaired by the JFES vice president, Uchida Moriya), and the Secretariat (chaired by the JSEE secretary-general, Harada Kosaku). Thereafter, the JABEE preparatory committee, chaired by Ohashi, was established in December 1998. Finally, JABEE was established in November 1999. JABEE made engineering ethics a requirement in the common criteria. Thus, engineering ethics education garnered much attention in Japan.

As the name of the JSEE–JFES joint committee indicates, JABEE was established primarily to ensure the training of engineers with international consistency. Ohashi explained at the committee's inaugural meeting that

> Globalization as a recent trend. We cannot live in isolation. We need a consistent system, transparency from the outside, quality assurance and constant improvement, and emphasis on individuality. Engineers' social responsibility is demanded, but human education is hollowed out.[6] (JABEE 2003, 11)

Emphasis on individuality as a unit of global economic deregulation and liberalization has heightened the importance of social responsibility. Moreover, despite the unclear reason, human education was considered hollow, and the emphasis on freedom and responsibility resulted in individualistic ethics education. Furthermore, the JABEE's founding prospectus rejected conventional Japanese engineering education because it "does not meet the requirements for developing engineers who act autonomously as members of a professional group working responsibly in society" (JABEE 2003, 36).

Hence, Ohashi conducted a study titled "Research on the Development of Evaluation and Accreditation System Aiming at the Improvement of the Quality of Engineering Education Program" funded by KAKENHI between 1997 and 1999, in which he established a trial version of the JABEE's accreditation criteria. Common Criterion 2 (Educational Outcomes) defined engineering ethics as "(a) an ability to consider what happiness and welfare of human beings are and related knowledge such as history, culture, etc." and "(b) an ability to consider and understand the effects and benefits of engineering solutions on society and the environment, as well as an awareness of ethical responsibilities as an engineer (engineering ethics)" (JABEE 2000, 13). Criterion (a) was revised to "an ability of multidimensional thinking with knowledge from global perspective," following criticism that happiness was subjective; thus, it was inappropriate as an educational criterion. With the strong demand for economic globalization and environmental sustainability, the criteria emphasized social responsibility in the same global perspective (JABEE 2012, 37–39; Ohnaka 2000, 23).[7]

When preparations for the establishment of JABEE began, ABET published Engineering Criteria 2000 (EC2000) in November 1996. EC2000 (adopted in 1997) shifted the emphasis from input measures to student outcomes, increased the curriculum structure flexibility, and required evidence of the professional preparation of graduates (Phillips et al. 2000, 98). In Japan, EC2000 was treated as a global standard of engineering education reform. From its first meeting in 1996, the JSEE Committee made repeated references to EC2000 (JABEE 2003, 11–49).[8] The following aspects of EC2000 are related to engineering ethics education: "(f) an understanding of professional and ethical responsibility" and "(h) the broad education necessary to understand the impact of engineering solutions in a global and societal context" in Criterion 3, and engineering design education considering "economic, environmental, sustainability, manufacturability, ethical, health and safety, social, and political" issues required in Criterion 4. Emphasizing engineering ethics regarding ABET's criteria, JABEE integrated points (f) and (h) of Criterion 3 in EC2000 into a single criterion (b) in its criteria.

Although its criteria provided a basic guide, JABEE wanted engineering colleges to be accountable for their education and never specified the curricula. Ohnaka, who became the chairperson of the JABEE's Criteria and Evaluation Committee, did not restrict the contents of engineering ethics education when explaining the criteria but illustrated how to teach and evaluate engineering ethics in approximately 260 words with simple examples (Ohnaka 2000, 24). Skeptical about introducing a new class on engineering ethics (one that JABEE did not request), Ohnaka (2004, 308) recommended the introduction of ethics across the curriculum by incorporating it into project-based learning classes through case studies.

As the education direction was abstract and ambiguous, many professors were confused by JABEE's decision to make ethics compulsory. Many study groups, workshops, and feature articles were organized to explore the teaching contents in the 2000s. Engineering professors, engineers (especially JCEA members), and scholars of science and technology studies (especially philosophy, ethics, and the history of science and technology)—who had little relationship with engineering professors before entering the field—collaborated in developing the teaching materials as an opportunity to increase the visibility of their fields. Thus, the introduction of engineering ethics education progressed through the combination of various interests.

Across the cases of the IPSJ, the SCJ, the JFES, the JSEE, and JABEE, the most significant driving force for the introduction of engineering ethics was compliance with global standards required by global communities. Engineering ethics was introduced to clarify autonomy and responsibility as a means of quality assurance for individuals to promote the global deregulation and liberalization of professional services. The need for human resources was represented in JABEE's catchphrase: "From products certification to human accreditation" (JABEE 1999, 2003, 42, 58). The introduction of engineering ethics, thus, prepared for the service-oriented industrial structure of the knowledge-based economy. Meanwhile, a growing tendency to seek individualistic ethics in Japanese society evolved since the mid-1990s. The following section discusses the domestic motivation for promoting engineering ethics education other than compliance with foreign standards.

EXPANSION OF ECONOMIC IDEOLOGY

The aim of engineering education reform changed drastically circa 1995. Where reform once had sought to meet the demand for engineering human resources in the domestic industry, it shifted to the realization of the international mobility of human resources in the global market. Nevertheless, both objectives formed part of Japan's industrial policy to restructure the economy into a knowledge-based economy, which had been promoted since the 1980s. Thus, reform was promoted through an industry-academia-government collaboration led by engineering professors with shared interests.

According to Yoshikawa, the first president of JABEE, "JABEE is one of the few cases in which a system with a nationwide vision was built through private-sector initiative amid Japan's common culture of government initiatives." He continued,

> What can we do with the power of the private sector driven only by the volunteer spirit? What should we do to promote our activities, not being self-righteous, but cooperating with the national policy with the understanding and support from the industry? We are proud of the JABEE's birth history, which contains many hints. (JABEE 2003, under "forward")

The founders of JABEE interpreted the external pressure not as an unfortunate requirement of globalization but as an original issue in Japan that could be resolved democratically. Their attitude was not only similar to the democratic approach of engineering professors in the engineering education reform following the defeat in World War II but also compatible with the idea of economic liberalization and free-market capitalism (so-called neoliberalism). Therefore, the reform was also actively promoted by the government and industry as a national policy.

The JSEE Committee meetings were observed by government officials from the MOE Higher Education Bureau in charge of *refresh education* since their first meeting in 1996. The accreditation system was promoted based on the relationship between the MOE and the JSEE in *refresh education* in the early 1990s, as part of vocational education. The JSEE-JFES joint committee meetings were similarly observed by government officials from the STA, which had jurisdiction over the Gijutsushi system, and the MITI. The observers from the STA and the MITI repeatedly commented on the APEC/HRDWG and the APEC Engineer agreement. As the director of the STA Science and Technology Information Division stated, "we would like to work together by forming an all-Japan alliance." In fact, regarding the all-Japan cooperation on the government side, several ministries (i.e., STA; Environment Agency; Ministry of Justice; Ministry of Foreign Affairs; MOE; MOHW; Ministry of Agriculture, Forestry, and Fisheries; MITI; Ministry of Transport; MOPT; Ministry of Labor; Ministry of Construction) jointly established the Ministry's Liaison Committee on the APEC Engineer Project (APEC Gijutsusha Shikaku Sōgo Shōnin Project Kankei Shōchō Renrakukai) in January 1999 (JABEE 2003, 18, 35).

The establishment of JABEE was promoted through industry-academia-government collaborations. Accordingly, Keidanren (the Business Federation) and the supreme adviser of the Tokyo Electric Power Company (TEPCO) attended the preparatory meeting of the JSEE-JFES joint committee in June 1997. Moreover, 102 people from the Japanese industry, academia, and government attended the inaugural meeting in July, including members of the Association of Principals of Technical Colleges, professional engineering associations (such as the JTTAS, the JCEA, and the AJCE), and seven newspaper companies. The vice president of Keidanren was among the founders of JABEE (JABEE 2003, 68).

Given that the establishment of JABEE was promoted as part of industrial policy, its philosophy had a strong affinity with the Japanese industry. In March 1995, the MOE established the industry-academia meeting to develop leading and creative human resources in science and engineering colleges and realize a "science and technology-oriented nation" (kagaku gijutsu rikkoku).[9] It also published an urgent proposal in July. In March 1996, Keidanren published a proposal, "Developing Japan's Creative Human Resources: An Action Agenda for Reform in Education and Corporate Conduct," based on the MOE proposal. The JABEE preparatory committee recognized Keidanren's proposal as a request from the industry (JABEE 2003, 54). In

January 1996, Keidanren announced its policy agenda, *An Attractive Japan: Keidanren's Vision for 2020.* Arguing for the need to strengthen the free trade regime, Keidanren declared:

> In particular, to be a leader in the liberalization of the APEC region, our country will decisively deregulate in accordance with the Osaka Action Agenda in November 1995, based on the policy of freedom in principle and regulation in exceptional cases. Moreover, our country will take the initiative in fully implementing the Bogor Declaration.[10] (Keidanren 1996a)

In this regard, the "promotion of deregulation in education" was sought to "establish a human-resources development system to foster creativity" (Keidanren 1996a). Keidanren's March proposal, "Developing Japan's Creative Human Resources," was in line with Keidanren's Vision for 2020.[11]

Furthermore, maintaining that "Japanese society will need human resources consisting of creative individuals who act independently and possess a strong sense of individual responsibility," Keidanren's March proposal called for the creativity of free-thinking individuals to resolve problems and emphasized the idea of *individual responsibility* (jiko sekinin) regarding freedom of choice (Keidanren 1996b). Keidanren further maintained that establishing individual responsibility would lead to "cultivating social respect for people's independence and raising awareness of social norms and ethics" because the existing social systems were at a complete standstill, inhibiting individual creativity. Moreover, "there is a need to fundamentally revise the educational system, corporate behavior, and organizational culture." The independence and individual responsibility advocated by Keidanren in the proposal are closely related to the economic liberalism of deregulation. Hence, liberalism was expected to be a basic principle for ameliorating various social problems beyond economic activities.

Neoliberalism—the belief in the power of the free market to optimize society—has long been at the center of Keidanren's ideology. Shortly after joining the bureau in 1963, Uchida Kozo, senior managing director of Keidanren, read Milton Friedman's *Capitalism and Freedom* and was struck by the idea that everything, even education, would work better if privatized and left to the market-based competitive mechanism. According to Uchida,

> Friedman's economic theory of the free market system, of the so-called Chicago school, is the most appropriate ideology and source of inspiration for Keidanren. Through more than thirty years of my service, I have the impression that his philosophy was basically sufficient to address the issues I was in charge of. Recently, deregulation has emerged as the most important issue facing Japan.

Its philosophy can be said to be the very essence of Friedman's philosophy. (Uchida 1996, 209)

Uchida studied Marxist economics as both an undergraduate and graduate student.[12] In reaction to his studies, he advocated for neoliberalism and supported Keidanren's activities. In the 1990s, as the Japanese economy stagnated due to a shift in the industrial structure alongside the global economy's expansion, Keidanren began proactively releasing proposals for social improvement. These proposals advanced the neoliberal ideology more clearly and argued for the liberalization of all aspects of society.

Liberalism and individual responsibility were not only emphasized by Keidanren during this period but also throughout the business community. Keizai Doyukai, the Japan Association of Corporate Executives, called for economic liberalization and *self-responsibility* (jiko sekinin, individual responsibility) in the Manifest for a Market-Oriented Economy in 1997. Criticizing conformism (yoko narabi taishitsu, follow-the-others mentality) as an improper corporate culture in Japan, Keizai Doyukai declared:

> Companies must break free of the conformism, which exerts strong control over Japanese corporate behavior and leads to inefficiency of the market as a whole. In essence, this conformism amounts to an evasion of responsibility, and the principle of self-responsibility will not be firmly established until it is eradicated. (Keizai Doyukai 1997, 10)

As part of the principle of *self-responsibility*, they expected companies to exhibit a new ethic and self-discipline based on the market principle, thereby eschewing the existing culture. Keizai Doyukai has maintained corporate social responsibility and publicness since 1955, adopting the principles of corporate management independence and individual responsibility. They developed activities to raise awareness by forming partnerships with the U.S. Committee for Economic Development (CED) in 1961 and publishing a translation of the CED report, *Social Responsibilities of Business Corporations*, in 1971 (Okazaki et al. 1996). The postwar business community consistently insisted on freedom and individual responsibility. However, in the 1990s, emphasis on individualistic responsibility resulted in the philosophy of economic activities being considered as though they were the ethics of society.

Neoliberalism was also strongly promoted by the government advisory board. In the July 1999 report, "Ideal Socio-economy and Policies for Economic Rebirth," the Economic Council (1999) stated that products and services, employment and work patterns, and people's attributions were diversified, employed "the maximum freedom and the least dissatisfaction" as the goals of the knowledge society with diversity, and emphasized the concept of freedom and publicness (*kō*) as principles of individual responsibility. Thus, the "ideal" was no longer understood as restricted to economic society but expanded to general society; alternatively, it may have indicated the understanding of general society as an economic society. The Economic Council argued that innovation through competition was necessary to realize an ideal society—a goal for which the existing values of efficiency, equality, and safety were insufficient. The report thus advanced the need for a new value of freedom: "justice." According to the report, this value was integral to ensuring solidarity among individuals, improving productivity, and sustaining economic growth through intense competition. The council asserted that Japan could not remain a major world economy player if the value remains unclear.[13]

As noted, individualistic ethics were emphasized to promote economic liberalism in a positive manner. Such ethics were even agreeable to those who took a negative view of economic liberalism, such as Murakami, who located the root of unethical behavior in free competition. Moreover, as discussed in chapter 5, corporate scandals have become a major issue in Japan since the late 1980s, stimulating public demands for corporate ethics.

In the earliest case, the October 1996 issue of *Total Quality Management*, a magazine published by the JUSE, featured corporate morals and quality. The preface claimed that the numerous corporate scandals involving securities, banks, general constructors, food, electronics manufacturers, trading, department, and pharmaceutical companies continued as an aftermath of the economic bubble burst in the early 1990s (JUSE 1996, 9). Fudano Jun's article, "Engineering Ethics Education in the United States," was presented as the first article of the special issue. As an engineering magazine, it would have made sense to focus on engineering ethics education. However, because the special issue covered a wide range of corporate scandals beyond engineering, it is unclear how effective the American engineering ethics education is as a direct measure. Fudano concluded that while the effectiveness of such education was not evident, it was unacceptable to do nothing; individual engineers had no choice but to do their best, having a strong sense of vocation and high ethical standards in aspiring toward a "science and technology-oriented creative nation" (Fudano 1996, 15). Despite the uncertainty of its effectiveness, suspicion regarding the existing corporate culture offered no choice but to call for new professional ethics emphasizing individuality.

In comparing American and Japanese engineering ethics, Americans are typically considered to possess an ethical principle of acting appropriately by ceasing unethical relationships; the Japanese are considered to possess ethics that depend on existing personal relationships (EAJ 2000, 8–9). Given that corporate scandals are attributed to the traditional ethics of Japanese collectivism and conformism, it seemed natural to focus on American individualistic ethics to break away from traditional culture. The argument was also compatible with the Japanese industry and government agenda to promote economic deregulation. Consequently, individualistic ethics became a particularly persuasive argument in the 1990s. Thus, the industry further strengthened both liberalism and individual responsibility, which had long been advocated by the business community, with the general public's acceptance. The next section analyzes this new trend in public opinion.

CRITICISM OF JAPANESE ORGANIZATIONAL CULTURE

Figure 7.1 shows the number of newspaper articles that mentioned both "ethics" (*rinri*) and "science and technology" (*kagaku gijutsu*), "engineering" (*kōgaku*), or "scientist or engineer" (*kagakusha, gijutsusha*) in major Japanese newspapers by year.[14] The primary axis shows the number of articles that refer to science, technology, and engineering, while the secondary axis shows the number of articles that refer to scientists and engineers; the digit of each axis is different. "Science and technology" and "scientist

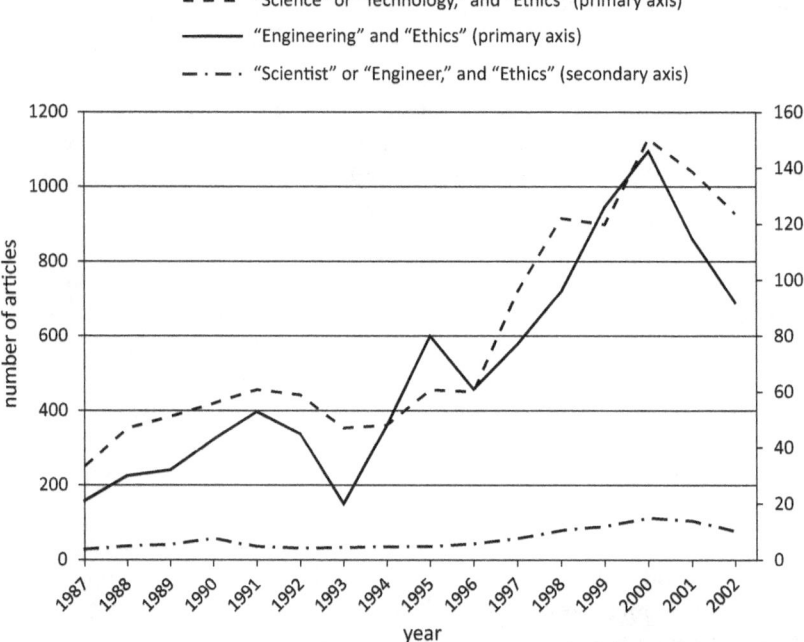

Figure 7.1 Number of Articles on both "Ethics" (*rinri*) and "Science and Technology" (*kagaku gijutsu*), "Engineering" (*kōgaku*), or "Scientist or Engineer" (*kagakusha, gijutsusha*) in Major Japanese Newspapers by year. *Source*: Data from Natsume (2015b, 56).

and engineer" saw a small rise and decline in the early 1990s followed by a significant increase toward the late 1990s.[15]

Serious incidents in 1995, such as the Aum Shinrikyo cult's sarin attack on the Tokyo subway and the concealment of the sodium leak accident at the Monju nuclear reactor, were later cited as the main reasons for increasing the appeal of engineering ethics. However, attributing these incidents to the ethics of scientists and engineers was not initially widespread in society. While the number of articles referring to scientists' ethics, even citing the cult Aum, increased in 1995, the dominant topic in ethics was the fact that it was the fiftieth year since the dropping of atomic bombs and Japan's defeat in World War II. Since 1995 constituted a historical milestone, there was a significant increase in the number of articles on postwar Japanese society. Moreover, the issue at hand was ethics for scientists, not engineers. In contrast, JABEE had little to do with the increase in the number of articles on engineers and ethics in 1999. Rather, the JCO criticality accident at the end of September influenced the coverage increase, with criticism of the ethics of engineers in the nuclear sector becoming prominent after the falsification of manufacturing data for nuclear fuel containers was revealed in 1998.

During the 1990s, the controversy over brain death in the early 1990s, the establishment of the Act on Organ Transplantation in 1997, and the controversy over cloning technology in around 1997 became major issues in biological and medical ethics. Combined with the influence of corporate scandals, public concern regarding ethics and morality increased slightly (drastically and cumulatively) in the early (late) 1990s (Natsume 2015b).

As figure 7.1 shows, criticism of the ethics of individual professionals appears to have increased gradually with the spread of public concern in the late 1990s rather than as a direct result of the 1995 accidents. The HIV-tainted blood scandal constituted a major turning point. Infected patients and bereaved families filed lawsuits for damages in 1989, claiming that many hemophiliacs were infected with HIV from unheated blood products in the mid-1980s due to predictable omissions by the MOHW and the pharmaceutical companies. The media began reporting on the issue in 1994. Sakurai Yoshiko (1994) accused Abe Takeshi, a medical professor who was the head of the MOHW's AIDS Study Group, of individual culpability in her award-winning nonfiction book on the issue. The lawsuit evolved into a citizens' movement in 1995, while Kobayashi Yoshinori (1996) defended the movement and condemned the professional ethics of Abe and the MOHW in his influential manga (comic) essay. Consequently, in February 1996, the MOHW minister, Kan Naoto,[16] met with 200 plaintiffs to acknowledge the legal liability of the government and apologize. The four pharmaceutical companies also accepted the settlement and apologized to the plaintiffs in March 1996.

Given reconstruction efforts following the Great Hanshin-Awaji Earthquake, 1995 became known as the first year of volunteerism in Japan. Expectations for the citizens' movement led to the enactment of the NPO Law in 1998. At the time, the basic tenets of the citizens' movement were democracy against bureaucracy, the importance of accountability, and individual power over the existing power, as preserved by the Japanese traditional culture of harmony but indifference (Wolferen 1994). However, Kobayashi broke away from the citizens' movement following the HIV lawsuits because, like the student activism in the late 1960s, the continuation of the movement itself became its purpose. Thus, he argued that priority should be given to establishing the independent individuality of each person rather than solidarity. The aim advanced in these arguments differed from that of the business community, as discussed in the previous section. Moreover, the citizens' movement was of a different opinion. However, they shared a common policy expectation of individual power in public.

Thus, professional ethics emerged as a social issue around 1996, but the general public did not immediately claim ethics for scientists and engineers. For example, the May 1996 issue of *Gendai shisō* (Contemporary Thought), a leading commercial magazine for humanities, published a special feature, "Who Is the Scientist?", inspired by the title of Murakami's book. Although the purpose of this feature was not explained, the editorial postscript included the following statement:

> Where is the scientist? This question emerged in the 1990s. Did we see scientists in the accident of the Monju fast breeder reactor? We have certainly seen many government officials, but do you remember the names of the scientists who were supposed to be among them? In the HIV-tainted blood scandal, we have seen government officials with the title of scientist or scientists with the face of government officials. Even so, we are also required to be scientists in the cases of informed consent, brain death, energy affairs, environmental pollution, and so on. So, the question [. . .] is not what science is, but exactly who the scientist is. (Ikegami 1996, 342)

While Murakami asked, "What is the scientist?," this special issue asked, "Who is the scientist?" in pointing to the challenge of identifying the responsibility of individual scientists who constituted an ambiguous group. Despite discussing individuals in organizations and fraud among scientists, the articles did not clarify whether they were advocating for professional ethics for individual scientists and engineers.[17]

While professional ethics was not a major issue in *Gendai shisō*, the second edition of *On Being a Scientist: Responsible Conduct in Research*, edited by the NAS, the NAE, and the Institute of Medicine on scientific ethics, was translated into Japanese in 1996.[18] The ethical norms this book addressed were mostly internal to the scientific community rather than external to the general public. However, in the afterword, Ikeuchi Satoru, the translator and the astrophysics professor at Osaka University, wrote:

> In 1995, many accidents occurred in Japan, such as the Great Hanshin-Awaji Earthquake, the Aum turmoil, and the accident of the Monju fast breeder reactor. In each case, the current state of science or scientists was brought into focus. The reality of earthquake prediction research, the collapse of highways that were guaranteed to be safe, the mind control of young people who were trained in science, and the accident of nuclear reactor, which was supposed to be the height of technology, have intensified people's skepticism about science and scientists. This happened because the forefront of science has moved away from the sense of daily life, and the almost arrogant mind of the scientists as the experts was apparent. (Ikeuchi 1996, 87)

Ikeuchi noted that Japanese scientists faced ethical problems, including those related to the HIV-tainted blood scandal and issues of organ transplantation, euthanasia, and environmental pollution. However, these issues had not become serious enough to damage the entire science sector because, unlike in the United States, Japanese scientists tended to keep details of problems vague because of the *vertical organization* (*tate shakai*) of their closed culture (Ikeuchi 1996, 90). Noting the need to change this culture, he argued that "we are entering an era in which the ethics of scientists are truly questioned" (Ikeuchi 1996, 90), largely due to the strengthening of graduate schools and research budgets under the "science and technology-oriented nation" policy based on the 1995 Basic Act on Science and Technology.

Nakane Chie, a social anthropologist, developed the concept of *vertical organization* in analyzing Japanese social structure. According to Nakane (1970), in Japanese society, homogeneous groups distinguished by the consciousness of *uchi* (our people) and *soto* (outsiders) form a *vertical organization* as a *social frame* based on patriarchal and bureaucratic ranking; members are expected to exhibit total emotional participation in their organization as moral actors. This organizational culture makes it easier for employees to choose behaviors that are close to the interests of their own company rather than the public interest. In the 1990s, when the *vertical organization* of Japanese culture became an issue, engineering ethics was expected to be free and autonomous from organizational constraints.

The major incidents of 1995 provided a significant opportunity for Japanese society to reconsider the relationship between individuals and the public, particularly insofar as they had a tremendous social impact as a series of events rather than respective isolated incidents. The growing momentum to reexamine Japan's postwar democratic system fifty years after World War II also contributed to the reconsideration. Thus, the questioning of the professional ethics of individuals became a common issue in Japanese society.

Elected as the SCJ president for the council's seventeenth term, Yoshikawa delivered his inaugural address, "Ethics for Scientists," in July 1997 (Yoshikawa 1997). Based on his experience of investigating the Monju accident, Yoshikawa identified the "Japan disease" as a problem in his "New Year's Opinion" in 1998:

> It is not clear who is really responsible for the accident and its countermeasures. It does not matter who is formally responsible. There are many things to do, from the technical practices on-site to the decision making of the corporation. However, the problem is the ambiguity of the person who has discretion over each of them and holds the responsibility for the result in reality, aside from the laws. (Yoshikawa 1998, 5–6)

Yoshikawa's argument was similar to that of *Gendai shisō*'s "Who is the scientist?," although he expanded the subject to include the ethics of engineers. Noting that "Japan has no common religion to support our basic ethics, as many other countries do" (Yoshikawa 1998, 7), Yoshikawa advocated scientific knowledge and the SCJ (its representative organization in Japan) to play the role of religion.

Although arguments for emphasizing individualistic engineering ethics occasionally emerged before the 1990s, they were not popular, as it concerned scientists more than engineers. Nevertheless, Murakami's 1994 book anticipated the issue of engineering education reforms in the late 1990s—partly because of the author's membership to RCAST, an engineering organization. In another pioneering case, Furuya Keiichi provided a comprehensive introduction to American engineering ethics in a general course, "Technology Studies," at the Faculty of Engineering at the University of Tokyo from 1987 to 1994 (Furuya 2003, A13).[19] The lecture was based on *Environmental Ethics for Engineers* by Alastair S. Gunn and P. Aarne Vesilind (1986). He further published the Japanese translation with Japanese cases in 1993. However, Furuya's efforts were an exception rather than the norm, which ended in 1994.

In another case before 1995, the Japan Society of Mechanical Engineers (JSME) launched the Technology and Society Division in 1990 to respond to the growing interest in technology ethics. At the 1994 Annual Conference, Fujie Kunio of ShinMaywa Industries Ltd. proposed *Gidō* (the Technology Way), as a moral theory of technique, in contrast with *gijutsu* (technology), the methodology of technique. According to Fujie, the harmonious coexistence between humans and nature had long been part of *Gidō*, in much the same manner mental discipline was required in *judo*, the Japanese martial arts. From this perspective and the environmental ethics of harmony with nature against machine civilization, Fujie (1994a) called for the conceptual shift from *gijutsu* to *Gidō*. Noting that, "in general, ethics is reminiscent of prewar moral training, which some people feel some kind of archaism and sometimes even reactionary toward," Fujie (1994b) maintained:

> In the fifty years after the war, Japan has become one of the advanced countries in terms of materiality, but there has been a period of blankness in social ethics in terms of mentality, and an era without principles continues. In recent years, the ethics of science and technology have become a major issue due to the emergence of various types of pollution, including global environmental problems. However, today's ethical issues are important not only for science and technology but also for politics, economics, and society in general. Among them, scientists and engineers must be aware of and act on their responsibilities and obligations to our civilization from the perspective of ethics for future sustainable development. (Fujie 1994b)

Fujie's emphasis on environmental conservation and sustainable development in advancing *Gidō* was rooted in the growing global concern for environmental issues. At the United Nations Conference on Environment and Development in 1992—also known as the Earth Summit—the Rio Declaration was adopted to establish principles for sustainable development. In 1993, Japan enacted the Basic Environment Act, which set out the basic policies for environmental conservation. Thus, Fujie's argument in 1994 can be considered an extension of the technology ethics of the 1980s, as discussed in chapter 5.

Although Fujie's original conception of *Gidō* focused on environmental ethics, it was developed as a form of general engineering ethics per the expanding nationwide interest in professional ethics. In his presentation at the JSME Annual Conference in September 1996, Fujie located the origin of his idea in the history of the Jin dynasty in ancient China, provided a Japanese case in the form of an employee training lecture given by Yokoyama Kozo at Mitsubishi Shipbuilding Co., Ltd. in 1952, and finally introduced the ASME's code of ethics adopted in 1975 (Fujie 1996, 434). Fujie's presentation was included in an organized session, "Ethics and Responsibility of Engineers," by the Technology, Society, and Engineering Education Division of the Annual Conference. In the session, Fudano of KIT presented on the current situation of engineering ethics education in the United States, including ABET's EC2000, the history of the learned profession, and the training of moral autonomy using the case method. The JSME subsequently organized the Engineering Ethics Study Group in the Technology and Society Division in January 1998. In October 1998, the JSME established the Arrangement Working Group of the Code of Ethics, which included Fujie and Fudano among its members, and approved a code of ethics in December 1999.

While several engineering professors and engineers were enthusiastic about the introduction of engineering ethics, others were bewildered by the move, considering it as a Western, particularly American, culture. The discussions at the "Reform of Engineering Education and Potential of STS" symposium organized by STS Network Japan—a citizens' academic society for science, technology, and society (STS)—revealed the diverse understandings of engineering ethics in 1999. As the symposium's first speaker, Fudano gave a presentation on the "New Trend in Engineering Education: An Image of Engineer in the Twenty-First Century and JABEE," in which he argued for the need for individualistic engineering ethics to ensure engineers make value judgments independently, particularly insofar as engineers tended to prioritize their companies' interests over those of the general public (Miyake 2000).[20] In contrast, the second speaker, Shibata Kiyoshi, an engineer at Nippon Steel Corporation, gave a presentation titled "Can Engineers Get out of Slavery?: Possibility of Engineers' Autonomy in Company." He described the limitation of attributing problems to individualistic ethics and value judgments, given the fact that engineers' specializations shifted as management assigned tasks. The third speaker, Kihara Hidetoshi, professor of science and technology studies, argued that the politics of scientists and engineers were more important than their ethics. Thus, he voiced concerns over the increasing domination of science and technology experts in making value judgments autonomously and raised questions about whether it was possible and appropriate for engineers to have such freedom and power. Meanwhile, Kihara also acknowledged the significance of engineering ethics in STS, noting the importance of focusing on individualistic ethics because sociologists tend to seek the causes of problems in abstract societies and organizations (STS Network Japan 2002). Moreover, voicing a different perspective, the symposium commentator maintained that engineering education reform was really a "business opportunity" for STS (Miyake 2000, 3). This argument reflects an aspect of the interest in promoting the introduction of engineering ethics in the late 1990s. While this symposium is not representative of the actual understanding of engineering ethics at the time, it reveals the confusion and expectation surrounding the issue.

AMERICAN ENGINEERING ETHICS EDUCATION AT KIT

The introduction of engineering ethics education has made substantial progress in Japan since the establishment of JABEE in 1999. Employing several full-time faculty members to offer "Science and Engineering Ethics" as a compulsory course for more than 1,600 students across its faculties each year, Kanazawa Institute of Technology (KIT) became the most influential university in engineering ethics education in Japan.[21] Having paid attention to American engineering ethics since the mid-1990s, Fudano Jun, professor at KIT, played a leading role in promoting engineering ethics education in Japan in the late 1990s, as described in this section.

The focus on engineering ethics was promoted by the education reform launched by KIT in the early 1990s, which was modeled on American engineering education. As with other universities, KIT's major issues included the SEU's generalization in 1991 and the peak in the eighteen-year-old population in 1992. KIT initially paid attention to American colleges to improve their research system. However, after surveying the United States, they found refuge in the American education system of engineering design rather than the research system. Under the new policy, KIT formed the Study Committee on Education Reform in 1992 and dispatched about half its faculty members (about 100 professors) and related staff to various universities in the United States to conduct a large-scale survey on engineering design education.[22] Thus, American engineering education caught the attention of KIT in the early 1990s.

While KIT's main interest lay in engineering design education and portfolio education, Fudano, director of the Office of International Programs, focused on engineering ethics education as a characteristic of American engineering education. After receiving a master's degree in science education from the International Christian University in Tokyo, Fudano received his PhD from the University of Oklahoma in 1990 for his research on the history of American X-ray science. Upon returning to Japan, Fudano began working at KIT and became the director of the newly established Office of International Programs in 1992. The fact that he studied at an American university, was fluent in English, and even specialized in science education and the history of science were important factors in his focus on American engineering ethics and his role in its dissemination.

In 1992, when Fudano became the director of the Office of International Programs and the institute's preparations for education reform began, KIT formed an international partnership with Rose–Hulman Institute of Technology (RHIT) and promoted frequent personal interactions and cooperation. KIT initially considered adopting ABET's accreditation and was permitted to read relevant documents when visiting RHIT. Although KIT eventually abandoned ABET's accreditation, Fudano and his colleagues became aware of engineering ethics education through RHIT.

Heinz C. Luegenbiehl, professor of philosophy and technology studies at RHIT, provided detailed information to Fudano. As a visiting professor at

KIT from March to August 1996, Luegenbiehl gave lectures on American engineering ethics at the graduate school. The Japanese title of the course was "Special Lectures on the History of Science and Technology" (Kagaku Gijutsushi Tokuron), but the English title was "Introduction to American Engineering Ethics" (Fudano 1999, 42–109).

The Science, Technology, Society, and Policy Program was launched at KIT in 1989 under the leadership of Tezuka Akira, a member of JSPS Committee 149. In 1991, the program obtained priority area funding from KAKENHI to conduct a four-year research project titled "A Transdisciplinary Study on Public Acceptance of Advanced Science and Technology in Modern Society." Fudano served as the principal investigator in 1993. Another Fudano research project, "A Study on Professional Ethics Education in the University Curriculum for Engineering and Science," was selected as a KAKENHI project in FY1994–1995. From December 1993 to January 1994, he conducted a questionnaire survey, investigating the understanding of his co-investigators, engineering professors, students, and Kanazawa citizens on professional ethics.

The survey asked: "Do you think that education on social responsibility and professional ethics is necessary in educating scientists and engineers?" According to the results, 40 percent of engineering students answered, "strongly agree," as did 60 percent of the other groups; when including "somewhat agree," 80 percent of engineering students agreed, as did 90 percent of the other groups (Fudano 1995, 23). Hence, Fudano concluded that the importance of professional ethics education was clear, and the general public did not appreciate the ethics of Japanese scientists and engineers. Accordingly, he advanced the need for the fundamental reform of professional ethics education in engineering education for Japan to play an active role as a respected leader in ethics in the twenty-first century (Fudano 1995, 51). Fudano later stated that the survey results convinced him that engineering ethics education would be widely accepted in Japan. Thus, in 1994, Fudano became acutely aware of the possibility of introducing American engineering ethics education to Japan.

American engineering ethics education was fundamentally different from the previous KIT education. KIT had advocated the "cultivation of high character" as well as "deep technical innovation" and "superb academic-industrial cooperation" as its founding principles since its establishment in 1965.[23] It also offered a history of science course as a compulsory subject for all first-year students since 1980 to cultivate a sense of ethics and mission in engineers. The course reflected the first chairperson's motto, "Don't be a devil with technology" (Chiku 1983a, xviii–xiv).

Although the course provided a basis for the development of engineering ethics education at KIT in a broad sense, the actual content of the lectures was standard in the history of science. The content was completely different from that of engineering ethics education in the United States. According to the course outline, "Only after observing science and technology as a whole, reviewing the path of difficulty and creativity of our predecessors, and establishing your own views of science, will you have your own sense of mission and passion for contributing to human society." However, the content of the course was structured around (1) the methods underlying modern science, (2) the historical development of science, (3) the relationship between the natural and social sciences, and (4) the future of science; there was no mention of ethics (KIT 1980, 86). The course was introduced in conjunction with the opening of the KIT's rare book collection, "Kōgaku no Akebono" (officially translated as "the dawn of science and technology," although the original Japanese title means "the dawn of engineering"), which comprised first editions from the history of science rather than engineering when the new university library was built in 1982.[24]

In contrast, the curriculum reform implemented in 1995 explicitly introduced the perspective of ethics into some lectures. The keyword "ethics" was found in three of the elective compulsory course outlines for 1995: Nature and Ethics, Science and Culture, and Science and Environment (KIT 1995, 143–145). Rather than the professors in charge of each class, these outlines were prepared and written by staff members who promoted education reform modeled on American engineering education. At first, the introduction of ethics was not fully understood on campus. Therefore, the introduction of engineering ethics was sought as a new alternative style of liberal arts education, which had been reduced due to the SEU's generalization.

From 1996, Fudano, a leading expert in overseas cases, actively promoted international exchange to disseminate the effectiveness of engineering ethics education in Japan. For example, he introduced American engineering ethics education in the special issue of the JUSE's journal in 1996. He also introduced it to the JSME and the SCJ's WFEO Subcommittee. Luegenbiehl also contributed to such dissemination, including presenting a lecture interpreted by Fudano on "Engineering Ethics Education in the U.S." at the EAJ in July 1996 (EAJ 1996).

In April 1997, KIT established the Applied Ethics Center for Engineering and Science (ACES), with Vice President Hori Yukio as the director, to systematically promote research on the ethics of engineering, science, and business. Hori was a member of the SCJ's WFEO Subcommittee and the person who introduced Fudano to the committee. The establishment of the ACES constituted an extension of Fudano's KAKENHI project on professional ethics in FY1994–1995 (Fudano 1999, 1–2). In March 1998, KIT placed an advertisement in the *Nikkei* newspaper with a large photograph of Fudano, titled "Engineering Ethics Education Increasingly Needed in the Age of Scientists and Engineers Making Decisions," to publicize their efforts in education reform.[25] The advertisement made a strong impression on JCEA members (Hatakeyama, interview by the author, August 31, 2012). Six months later, the JCEA published the first Japanese textbook on engineering ethics in September 1998 (the Japanese translation of *Engineering Ethics: Concepts and Cases* of Charles E. Harris Jr. et al., as described in chapter 6). Accordingly, KIT's third education reform, which began in 2004, introduced "Science and Engineering Ethics" as a compulsory subject for all 1,600 third-year students. Thus, KIT realized the most large-scale and distinctive case of engineering ethics education in Japan.

AUTONOMOUS REINTERPRETATION OF EXTRINSIC ETHICS

External factors such as the APEC Engineer Register played a major role in introducing engineering ethics education in Japan. However, because the APEC and JABEE institutionally demanded engineering ethics as a global standard, it was considered an extrinsic and heteronomous requirement by many engineers and professors who felt pressured to accept it. They wanted to decide for themselves how their field needed ethics education. As ethics is supposed to be autonomous and not controlled by outsiders, even though the introduction process was not, an independent and spontaneous interpretation was desired in the acceptance process. Feelings of antipathy or unfamiliarity resulted in arguments that Japanese engineers had been ethical in the past, and other areas, such as management, required ethics more than engineering. They felt the introduction of a new educational course on ethics was unnecessary, provided engineers continued to do their jobs properly. Accordingly, autonomy became a major issue in the subsequent phase of the institutional introduction of engineering ethics.

The change in interpretation is illustrated by the difference in the translators' preface to the first and second editions of the JCEA's translation of *Engineering Ethics: Concepts and Cases*. While the preface to the first edition did not mention any specific incidents, it emphasized the difference in perceptions between the United States and Japan, given the international competition, as follows:

> We tend to look at things from a narrow and limited perspective of our own expertise in science and technology. This book shows that these perspectives are necessary for engineers in Japan, the United States, and other countries to collaborate and compete in the world under global standards. If we do not change, the difference will become a threat to the future. (Umeda 1998, III)

Published in 2002, the preface to the second edition promoted ethics as a domestic issue. After referring to the September 11 terrorist attacks in the United States in 2001 and the cult Aum's sarin attack on the Tokyo subway in 1995 in reflecting on the religious judgment of right and wrong, the preface stated:

> What is required for technology now? What should be the behavior of engineers? The above cases are typical of the irrecoverable confusion caused by the misuse of modern technology by people who lack a sense of ethics; in other words, those who cannot judge right and wrong. Even a person with a normal mind often causes an accident due to a misconception or inadvertent mishandling. This is why safety management is strictly demanded at work sites. (Sato 2002, III)

Although written by different people,[26] these prefaces are reflective of perceptions at each period. After the publication of the Japanese translation of *Engineering Ethics: Concepts and Cases*, members of the JCEA/TTC organized ethics seminars, studied under Imamichi Tomonobu, an ethicist, and began translating another textbook on engineering ethics to deepen their understanding of engineering ethics.[27] Through such activities, various incidents in Japan were investigated, reinterpreted, and rediscovered from the engineering ethics perspective.

The transition is similarly evident in the difference between the first (2001) and second (2002) editions of another textbook, *University Lecture: Introduction to Ethics of Engineers*, written by Sugimoto Taiji and Taki Shigeatsu at the JCEA. In chapters 4 (Formation of Engineers as Professionals), 8 (Connection between Engineering Education and Ethics), and 9 (Development of Code of Ethics in Engineering Societies), Taki discussed the topics based on his own experiences of the introduction of engineering ethics in Japan, such as the introduction of the APEC Engineer system and the establishment of ethical codes of academic societies (Sugimoto and Taki 2001). However, the second edition explored a new direction, that is, "how to explain engineering ethics in Japan." Thus, it emphasized Japanese culture, laws, and case studies, and reduced the international affairs to the final chapter, chapter 15: International Relations of Engineers (Sugimoto and Taki 2002, i–ii).

A similar transition can also be found in the discussion on the amendment of the Gijutsushi Act (from the CE Act to the PE Act) by the Committee on Science and Technology of the House of Representatives. At the committee meeting on March 14, 2000, the STA secretary explained that the amendment was to ensure international consistency and further develop high-quality Gijutsushi. Regarding the APEC Engineer issue, the secretary stated that "with the expansion of international economic activities, there is an increasing need to promote cross-border activities of engineers." Moreover, the Gijutsushi system is "expected to contribute to the improvement of Japan's science and technology by providing applied and advanced professional ability." Furthermore, he briefly stated that engineers were required to secure public interests, such as public safety and environmental conservation, based on the requirement of the APEC Engineer Register (Committee on Science and Technology 2 (147th Diet Rec. Mar. 14, 2000)). In short, the amendment was primarily to ensure consistency between the Gijutsushi system and the APEC Engineer system.

However, the subsequent committee meeting on March 29, 2000, focused on developing professional engineering ethics to establish a safety culture rather than international consistency. To emphasize its importance, the committee referred to accidents between 1999 and 2000, including the criticality accident at the JCO's Tokaimura nuclear fuel-processing plant, rocket launch failures, the concrete spalling accident in the tunnel of the Sanyo Shinkansen line, and the derailment of the Hibiya subway line. Thus, at the start of the meeting, the STA secretary claimed that "the reliability of Japan's technological infrastructure has been severely compromised," and "it is very important to raise awareness and morals for safety throughout the entire society and create a so-called safety culture to take root in the society." Accordingly, participants recognized the common "lack of morality" among engineers in these accidents (Committee on Science and Technology 3 (147th Diet Rec. Mar. 29, 2000)). However, this occurred at the final stage of the amendment, with the incidents between 1999 and 2000 supplying supplementary justification. While the amendment primarily addressed the international problem of the APEC Engineer, it was interpreted and accepted as a measure to address the domestic problems regarding the need for professional ethics.

Engineering ethics education was introduced in response to the global trend in engineering qualifications. However, at the acceptance stage, its effectiveness was emphasized by addressing the social problems related to science and technology faced by Japanese society at the time. This situation was a natural step in accepting the new and foreign concept of engineering ethics. It was partly because autonomy and spontaneity were essential for ethics; moreover, the general public sought a new framework to address the successive incidents that had undermined confidence in Japan's technological infrastructure. Of course, these incidents had varying causes, and it remains unclear how much of a role a "lack of morality" played. Nevertheless, given the reinterpretation of engineering ethics from such interests, the original purpose of adapting to the global trend became secondary.[28]

Meanwhile, the declining importance of APEC in the Asia-Pacific region since circa 1998 weakened the importance of international affairs in engineering ethics education. After the Asian financial crisis of July 1997, China's accession to the WTO in 2001, and the progress in the ASEAN diplomacy, the Asian economy changed drastically in the 2000s. Moreover, Japan's refusal to deregulate its forestry and fishery industries during the 1998 APEC Early Voluntary Sectoral Liberalization (EVSL) talks and the U.S.' attempt to use APEC as a security framework regarding the September 11 terrorist attacks exacerbated the reluctance of APEC member countries. Japan then changed its trade policy, shifting from an emphasis on multilateralism to conclude bilateral FTAs and economic partnership agreements, beginning with Singapore in 2002. As discussed in chapter 6, the shift from bilateralism to multilateralism facilitated the introduction of engineering ethics in Japan. It is worth analyzing the impact of the change in trade policy and the decline of APEC's importance in the 2000s on Japan's engineering qualification and education system, including engineering ethics; however, twenty-first-century developments lie beyond the purview of this book.

Nevertheless, reflecting a notable trend that emerged in the twenty-first century, the fourth edition of *Engineering Ethics: Concepts and Cases* by Harris et al. (2009) introduced a new perspective of *aspirational ethics* in contrast to traditional *preventive ethics*, which Fudano and his JSEE colleagues promoted as "encouraging engineering ethics" (*genki no deru gijutsusha rinri*). This trend became even clearer after the 2011 Great East Japan Earthquake. The nuclear accidents and misconducts of the 1990s heightened public awareness of engineering ethics; however, the 2011 Fukushima Daiichi nuclear disaster further impacted this perception. Fudano, an external member of the TEPCO Group Corporate Ethics Committee since 2009, has repeatedly emphasized the importance of aspirational ethics, especially after the 2011 disaster (TEPCO 2002–2015). His advice is based on the recognition that "it is effective to share good practices within the company in order to encourage employees to have a positive mindset," and "it is important to take measures that will enable new staff to have a mindset to work with pride as employees of TEPCO" (TEPCO 2014). Those involved in the nuclear power industry and TEPCO in the wake of the Fukushima disaster required this vital perspective.

Hence, Fudano focused on the trend of *positive psychology* and developed a new direction of *positive engineering ethics* that reflected it. Although it aims to realize the *well-being* and *happiness* of engineers, it remains an abstract statement of the ideas. It lacks a critical analysis of the effect of psychological ideas on individual engineers and society. Furthermore, it is not based on a specific case analysis, such as the accidents in the 1990s and the 2000s, that induced discussions on the need for engineering ethics (Fudano 2015).

The findings of psychology are only an analysis of mental states, not social responsibility. As described in this study, engineering ethics was introduced to Japan due to social demand, and its main purpose was to obligate engineers to act per social norms and responsibilities. In the early 1990s, Fudano began promoting engineering ethics by assessing its demand as a general public opinion. However, it tended to limit engineers, because it was seen as a control from the outside. In other words, while the realization of ethics essentially requires autonomy, its dissemination has increased heteronomy. Thus, *aspirational ethics* or *positive engineering ethics* sought to improve or neutralize the negative mindset.

However, if the realization of positive psychological states in individual engineers becomes a goal, it could distract attention from the historical cases that stimulated the social demand for engineering ethics and caused negative emotions. It could also widen the gap between the professional communities and the general public. The educational policy to promote autonomy and positivity will not prevent accidents, despite being a primary social requirement. Thus, the content of *happiness* must be thoroughly scrutinized from a social perspective.

Given the historical process, a fundamental dilemma exists between autonomy (positive mindset) and social demand (obligation of prevention) in Japan. The resolution has become a major challenge today. Thus, to resolve this issue, it is necessary to accumulate well-balanced empirical knowledge based on the positive and detailed analysis of actual cases. It is also necessary to develop a new philosophy to mitigate the opposing views. The philosophy should be associated with individualism and social responsibility, democracy, economic ideology, domestic culture, and the social status of engineers. The answer lies in the society-wide values regarding the use and control of science and technology in the future.

NOTES

1. In addition, NAFTA signed a mutual recognition agreement for engineering profession in June 1996. The Eight-University Council of Deans of the Engineering Faculties also established the Study Committee on Educational Programs in Engineering in 1996, which examined the educational system with engineering design and outcomes assessment based on domestic and international surveys of core curricula. However, the committee did not consider an accreditation system.

2. According to Ohnaka, the decision to establish the committee was also motivated by external pressure from the APEC Engineer agreement, as highlighted by the common Japanese refrain: "never change without external pressure" (JABEE 2012, 22).

3. Stating the need to question technological systems in the context of diverse values, such as environmental ethics with regard to energy resources, the report suggested the introduction of perspectives of public safety, health, and welfare, as

well as environmental responsibility and sustainable development (SCJ 1997). The SCJ's June 1997 report emphasized global environmental issues on engineering ethics. WFEO adopted the Code of Environmental Ethics for Engineers in 1987 and planned to hold an international conference, "Engineering Education and Training for Sustainable Development," in September 1997. The third session of the Conference of the Parties to the United Nations Framework Convention on Climate Change (COP3) was scheduled to take place in Kyoto in December 1997.

4. In addition to ABET, the examples of the code of ethics were drawn from the articles of the ASCE, WFEO, the JSCE, and the JCEA, with a list of the names of foreign academic societies that had adopted the code of ethics.

5. Prior to the report, the SCJ's Section 5 proposed increasing public support to academic societies in May 1996; the proposal noted that the functions of academic societies included "establishing professional ethics as a standard of conduct" and increasing public recognition of the profession (SCJ 1996, 60). Although the report was based on Murakami Yoichiro's *What Is the Scientist?* (SCJ 1996, 59), its understanding actually centered on the 1980 SCJ Charter for Scientific Researchers, where ethics is mentioned in only one sentence.

6. Ohashi then explained the concept of profession and license systems in Western countries, maintaining that Japan lacked understanding of whether the quality assurance of higher education was sufficient for engineering education and arguing that it was necessary to popularize the engineering license system (JABEE 2003, 11).

7. Compared to Western standards, Downey et al. (2007, 480) found the influence of Japanese traditional familism in JABEE's criteria for prioritizing contribution to society over an ability in mathematics and science or engineering analysis. However, in Japanese engineering ethics in the late 1990s, it should be noted that individualistic ethics were rather expected to overcome the traditional mindset.

8. In the United Kingdom, EngC enacted the Standards and Routes to Registration (SARTOR) for the registration of professional engineers and accreditation of engineering education programs in 1990 and revised it to emphasize the evaluation of educational outcomes similar to ABET. The JSEE Committee also examined the final draft of SARTOR and the cases of the Canadian Council of Professional Engineers and the Institution of Engineers Australia. Thus, to ensure compatibility with the existing Japanese system, they decided that engineering education should follow the ABET system and engineering license registration should be entrusted to another organization (JABEE 2003, 14). However, some held different opinions regarding the relationship between educational accreditation and engineers' licensing. Uchida Moriya suggested that JABEE's Coordination Committee on Engineers' Licensing Issue should establish the Japan Association for Professional Engineers to function as the APEC Engineer Monitoring Committee (JABEE 2003, 18).

9. Enacted in November 1995, the Basic Act on Science and Technology provided a basic plan for the budget under the slogan "science and technology-oriented nation."

10. The declaration was adopted at the APEC leaders' meeting held in Bogor, Indonesia, in 1994, which set forth the Bogor Goals.

11. Under Toyoda Shoichiro, Keidanren reorganized and expanded the Corporate Ethics Committee in 1996, revised the Keidanren Charter of Corporate Behavior in December 1996, and prepared the Implementation Guidance.

12. Uchida studied Uno Kozo's Marxist economics, which had a strong influence in postwar Japan.

13. The report also defended the globalization of the Gijutsushi system (Economic Council 1999).

14. This study employed Nikkei Telecom 21 (http://t21.nikkei.co.jp/), a newspaper article database provided by Nikkei Digital Media, to count the articles in major newspapers, including *Nikkei Shimbun*, *Nikkei Business Dairy*, *Nikkei Marketing Journal*, *Asahi Shimbun*, *Mainichi Shimbun*, and *Yomiuri Shimbun*.

15. The small number of articles on "engineering" (*kōgaku*) is probably due to the use of the Japanese term "kōgaku," which is generally defined as an academic discipline. A large part of the meaning of "engineering" in English is included in the term for "technology" (*gijutsu*) in Japanese.

16. Kan became a politician as a citizen activist following his experience of student activism and founded the Democratic Party of Japan in 1996. After the merger, the Democratic Party took over from the Liberal Democratic Party in 2009, and Kan was the prime minister when the 2011 Great East Japan Earthquake struck.

17. The special issue generally focused on the causes of problems and countermeasures in the current state of society rather than ethics. The introductory dialogue concluded with a section, "Shift the Gear of Science Education: What Is the Scientist?," discussing the necessity of science education as *sentiment education*. However, it lacked perspective on ethical education for individual scientists (Sato and Takeuchi 1996, 66–68).

18. Published by the NAS in 1989, the first edition has not been translated into Japanese.

19. Shortly after the University of Tokyo (Tōdai) Riots, Furuya pointed out the need for engineering ethics in engineering education, albeit briefly (Furuya 1971, 35).

20. Fudano's presentation consisted of a summary of Murakami's argument, an explanation of the American engineering education system, and the major positive opinions regarding the introduction of engineering ethics. In 2004, Fudano succeeded Murakami as a member of the World Commission on the Ethics of Scientific Knowledge and Technology (COMEST), UNESCO.

21. According to the survey by the JFES Council on Ethics for Engineers, only 2.8 percent of all Japanese universities had more than 1,000 students who took engineering ethics in the same grade (Minagawa 2009). In 2009, KIT was a leading university, with seven full-time faculty members employed regarding engineering ethics, five of whom taught classes.

22. At the time, a national policy was implemented to temporarily increase student enrollment to address the peak in the eighteen-year-old population; the revenue produced by the increased tuition fees provided the financial allowance necessary for such a large-scale survey.

23. The 1963 Central Council for Education report repeatedly used the phrase "cultivation of character" (*ningen keisei*) as a principle, mission, and duty of universities. It was a common issue in Japan at the time.

24. In his commentary on the rare book collection, Chiku Kakugyo, director of the Subject Librarian Office, advocated "establishing the ethics of engineering engineers." However, he only referred to ethics to justify his historical perspective. Most of his commentary was simply an introduction to the rare books with specific descriptions of the achievements of Western scientists and had nothing to do with ethics. Moreover, his arguments concerned the study of the history of science rather than engineering (Chiku 1983b).

25. It was one of a series of advertisements featuring KIT's faculty members, each according to a theme.

26. Each preface was written by the JCEA president at the time (Umeda Masao for the first edition and Sato Kiyoshi for the second edition). The preface to the first edition was written by Umeda himself, while the preface to the second edition was drafted by Kashima Minoru of the JCEA/TTC (Fukumoto, Hamada, and Kurosawa, interview by the author, July 13, 2012).

27. For details of the activities, see the "Project Team Activities" reports in the JCEA journal. Imamichi was a one-year junior to Kashima of the JCEA/TTC at the First Higher School under the prewar system. As mentioned in chapter 5, Imamichi developed science and technology ethics and introduced unique concepts, such as *eco-ethica* and *technology-mediated environment* (*gijutsu renkan*).

28. For insight into engineering ethics in Japanese society around 2000, see IPEJ (2001a).

Conclusion

This study has discussed the process through which engineering ethics was introduced in Japan in the twentieth century. Three major opportunities led to its introduction: the prewar trend of Taisho Democracy, the postwar Allied Occupation, and economic globalization. The responses to these opportunities were modeled on Western (particularly American) engineering qualification and education systems regarding engineers' social status, which engineers and engineering professors voluntarily promoted.

The first stage of introducing engineering ethics to Japan was the JCES' establishment of the Creed and the Code of Practice of Civil Engineers in 1938, modeled after the ASCE of the United States. As part of the movement to improve the social status of engineers, the JCES attempted to shift from an academic society to a professional society and formulated a code of ethics to facilitate the shift. Nevertheless, the code of ethics was immediately forgotten against the rise of Japanese nationalism. Prewar engineering education initially emphasized vocational training; however, it gradually emphasized academic education. Although moral education was promoted in middle level and lower education, nationwide paternalistic familism was emphasized as moral, while specialized professional ethics was not. This emphasis was modeled on British gentlemanliness and Japanese Bushido. Whereas the cultivation of nationalism was required by law, universities provided no ethics classes due to their focus on academic programs. Moreover, many higher technical schools that provided vocational education were converted into universities. This situation reduced the opportunities for engineering ethics education.

The second stage was the engineering education and qualification reforms, promoted by the occupation policy of GHQ/SCAP for democratization and industrial reconstruction after World War II. Japanese high-level engineers that sought work and social status in postwar reconstruction established the JCEA in 1951, modeled themselves after the American CE profession, and formulated the Code of Service for CEs. However, the JCEA's code of ethics did not induce any educational activity nor become a model for other engineering societies beyond its enactment. The JCEA was aware of the American PE system but aimed to introduce the American CE profession to Japan. The postwar engineering education reform leaders also recognized the importance of engineering ethics and regarded *cooperation* as the most important value for engineers. However, followers were reluctant to introduce engineering ethics, thus, rendering it unpopular and forgettable during the subsequent period of economic growth.

Leaders of the postwar engineering education reform claimed that cooperation was important to Japanese society but inadequate in its current form. According to Johnson (1982, 8, 323), who analyzed Japan's industrial policy history, the Japanese people generally have a reputation for having a "special capacity to cooperate," which is considered an irreducible Japanese cultural trait. However, it is largely derived from the situational motivation for national survival rather than basic Japanese values. Although there was much cooperation in Japan, it was perceived as a social and economic mobilization rather than a voluntary act (as demanded by American democracy). After the occupation period, reform leaders could not obtain sufficient industrial cooperation. Furthermore, industry-academia cooperation became a taboo until the 1980s due to the desire for academic independence and an aversion to the prewar nationalistic cooperative system.

During the occupation period, the CCS staff members noted Japan's lack of a development concept. However, Japan's development systems progressed in the subsequent economic growth period. Consequently, insufficiency or lack of research result transparency in Japan's R&D activities became noticeable. In the 1980s, the promotion and institutionalization of basic research became a critical issue in Japanese industrial policy at a domestic and international (Japan-United States) level. This policy issue prompted the institutionalization of industry-academia cooperation and the reform of engineering education in Japan.

As this international environment was based on economic interests, the value of competition saw significant emphasis relative to previous democratic cooperation. This situation was not a trend only in Japan. In the United States, the character of democracy as social capital had altered since the 1970s, and personal networks in vocational communities, including professional associations, declined (Putnam 2000, chap. 5, 439, 444). Since the 1980s, the global expansion of neoliberalism in the economy, which strongly promotes free-market capitalism, has altered the character of democracy. Consequently, by the end of the twentieth century, the value of cooperation was emphasized less.

The third stage was reorganizing the Gijutsushi system into the Japanese PE system and introducing JABEE's accreditation for professional education. Both stemmed from economic globalization, which rendered engineering ethics compulsory as a new method that conformed to the global standard. In the 1990s, engineering ethics in Japan was introduced due to the pressure by IFIP and WFEO to establish an ethics code. APEC also applied pressure to reform education. The engineering education reform initiated by the JSEE and the JFES in the mid-1990s was part of Japan's national industrial policy to address the APEC Engineer issue, which regarded a mutually recognized agreement on engineering licenses. The relevant ministries and industries supported the establishment of engineering qualification and educational accreditation systems per the global standard, fearing that Japan would otherwise miss out on globalization and be disadvantaged. Gijutsushi engineers also changed their qualification from CE to PE and willingly introduced Western engineering ethics to comply with the global standard, which threatened their social status. Moreover, in response to the SEU's generalization in 1991, some professors considered engineering ethics an important new subject that could create significant new possibilities for vocational education and liberal arts education.

Academic culture saw gradual transformations as the twentieth century drew to a close, in which regard, economic liberalism had a significant impact. In the late 1990s, engineering ethics became a new individualistic method of human resource development, which clarified that engineers were individually responsible for their respective professional services, not their companies. It also ensured the global economic mobility of engineering services. In the new world order established after the Cold War, the capitalist economic globalization compelled Japan's industrial and academic activities to adapt to a knowledge-based economy. Simultaneously, corporate ethics became a social problem, especially as corporate scandals increased with the economic bubble bursting in the early 1990s. Moreover, various significant incidents, such as the Aum cult's sarin attack and the sodium leak accident at the Monju nuclear reactor, have occurred since 1995, exacerbating the problem. Furthermore, individual responsibility, freedom, and independence have become more general philosophies for improving social problems due to economic liberalism. For this economic ideology, individualistic engineering ethics were expected to be a new method to complement organizational-level corporate ethics. Hence, reflections on Japan's postwar democracy became an important social issue in 1995, fifty years after the defeat in World War II.

If not for this adjustment in public perception, technology studies and the history of technology (rather than engineering ethics) may have been proposed as an educational subject in the late 1990s, as in the late 1980s. In postwar Japan, scientists, science and technology studies researchers, and critics have continuously discussed the social responsibility of individual scientists, especially regarding nuclear development. Nevertheless, the social responsibility of individual engineers had rarely been discussed. Technology ethics is an academic study of ethical issues related to technology promoted by academic communities other than professional engineering societies, which were expected to promote ethical codes and education. In Japan, in the 1980s, applied ethics garnered increased attention in the environmental, life, and information sciences, while engineering ethics remained undeveloped. Private sector engineers were separate from the academic community because industry-academia cooperation was not pursued.

Thus, external pressure from global society was essential for the institutionalization of engineering ethics in Japan. However, this globalization had not been a primary issue in the practice of Japanese engineering ethics education. When engineering ethics became an institutional requirement through global standards, such as APEC and JABEE, it became an extrinsic and heteronomous pressure, which was inconsistent with the autonomy and spontaneity intrinsically required by ethics. Moreover, the Western concept of "profession" was unfamiliar to Japanese engineers, who belonged to a collectivist and conformist culture. Given the subject matter of ethics, it was desirable to reconfigure it into something autonomous in the quest for better education. Furthermore, a series of significant incidents undermined confidence in Japan's technological infrastructure during this period, and Japanese society began seeking new ideas to address these new domestic problems. Hence, the original aim of creating international consistency in engineering qualifications shifted, thereby reforming engineering ethics education to enhance autonomy and spontaneity. The importance of engineering ethics education is no longer in doubt (from the third stage of its introduction in Japan). However, there remains doubt over how actively engineers promote ethics education and the extent to which the general public expects it.

The above discussion demonstrates that the stages by which engineering ethics was introduced in Japan were always related to the profession's role in democracy. Various historical studies of the relationship between the ethics of *scientists* and postwar democracy have been published in Japan, but few studies have examined the ethics of *engineers*. Democracy is an implicit precondition in Western engineering ethics. Given the possibilities of engineering ethics development in different cultures, democracy may not be an indispensable value. In Japan, however, with the strong influence of Western culture (particularly the United States), engineering ethics was repeatedly expected and eventually institutionalized in the 1990s. The glowing attention it attained was sparked by democratic values. Thus, engineering ethics may need to be promoted autonomously as part of a democratic movement to improve the social status of engineers for them to be active in its wide institutionalization. Alternatively, it may need to be demanded authoritatively as part of the dedication of engineers to the nation-state. Even so, the latter is less likely than the former as it would be coercive and heteronomous. The above historical considerations rely on Japan's historical and cultural context, which is unique in the world. Nevertheless, for the people of every country, these historical details provide thought-provoking examples of how to attain broader perspectives on the future management of science and technology.

Generally, autonomous and spontaneous management, rather than top-down management, seems to be indispensable in reducing science and technology risks as much as possible in highly advanced and complicated societies with a division of expertise. However, some significant questions remain. In the global economy consisting of nations with various political ideologies, how effective are the ideas of individualistic liberalism? How can it explain the relationship between professional ethics and the real causes of serious incidents related to science and technology? Moreover, how can it prevent such incidents? The answers to these questions will significantly influence the future of our society. The concept of publicness in Japanese society has been altered markedly by introducing Western concepts of democracy and individualism. Further historical analysis of the modifications to Japanese engineers' understanding of publicness before and after World War II will be required, along with an analysis of the causes of various incidents. This study provides a solid foundation for further consideration of these issues.

Supplemental Glossary

APEC Engineer

A common qualification for the mutual recognition of substantial equivalence of member countries' engineering qualifications to promote technology transfer within the Asia-Pacific Economic Cooperation (APEC) economies. Based on the APEC Osaka Action Agenda in 1995, the project was launched in 1996. Japan's major concern was whether to allow Gijutsushi to be a corresponding qualification in Japan. Engineering ethics had become a criterion for the selection. The Japan Accreditation Board for Engineering Education (JABEE) was founded as an accreditation body for engineering education to support this system.

Bushido (武士道, the Samurai Way)

A moral philosophy that developed among the Japanese samurai warrior class. It was systematically refined based on Confucian ideas during the Edo period. Moreover, it was reevaluated relative to Western ethics around the Sino- and Russo-Japanese Wars period and came to be regarded as a national morality based on the Japanese spirit. The term Bushido, hinging on the modern nationalistic view, was made famous by Nitobe Inazo's book, published in 1900.

Gendai shisō (現代思想, *Review of Contemporary Thought*)

A leading Japanese commercial magazine in the humanities, first published in 1973. Focusing particularly on French philosophical movements, such as structuralism and post-structuralism, the magazine increased its circulation with the French title, *Revue de la pensée d'aujourd'hui*. It features various monthly special topics on philosophical and social issues and has also focused on science and technology studies since its first issue.

Gijutsushi (技術士, the Japanese CE or PE qualification)

Japan's comprehensive national qualification for engineers specialized in various technical disciplines. The Japan Consulting Engineers Association (JCEA) was established in 1951 during the postwar reconstruction. Moreover, the Japanese CE Act was enacted in 1957, rendering it a national qualification. Initially, the system was designed to introduce consulting engineering services to Japan; however, after the amendment of the Act in 2000, it became a qualification for general professional engineers, not just for CEs.

Gijutsushi Fukumu Yōkō (技術士服務要綱, the Code of Service for CEs)

A code of ethics adopted by the JCEA in 1951. The notion behind its enactment and the content of each article came from the American Institute of Consulting Engineers and other U.S. codes of ethics.

Gijutsushi Gyōmu Rinri Yōkō (技術士業務倫理要綱, the Code of Ethics for CE Practice)

A code of ethics adopted by the JCEA in 1961. The ethical codes of the European CE associations were used as references in its formulation, especially on the issue of the independence and neutrality of the position.

Hirayama Fukujiro (平山復二郎, 1888–1962)

A civil engineer. After working for the Railways Agency and the Ministry of Railways, he became a director of the South Manchuria Railway in the prewar period. He was also a member of the Japan Civil Engineering Society (JCES) Investigation Committee for the Civil Engineers' Mutual Agreement. After the war, he led the early JCEA and became its president in 1959. He also served as the president of Pacific Consultants Co., Ltd., and the president of the JCES.

Imai Kaneichiro (今井兼一郎, 1917–2020)

A jet engine engineer and the board director of IHI Corporation. He also served as the president of the Japan Society of Mechanical Engineers. An international project alerted him of the need for a global engineering license.

Therefore, as the director of the Japan Technology Transfer Association, he promoted a plan to bring the American PE examination to Japan. From the 1980s to the mid-1990s, he eagerly worked to improve Japanese engineering education as the vice president of the Japanese Society for Engineering Education (JSEE) and the chairperson of the Engineering Academy of Japan (EAJ) Engineering Education Committee. Imai served on the World Federation of Engineering Organizations (WFEO) Committee on Education and Training and the U.S.-Japan Joint Task Force on Engineering Education.

Inose Hiroshi (猪瀬博, 1927–2000)

An electrical engineering professor at the University of Tokyo. He invented the time-slot interchange system for the digital telephone network and promoted the development of a traffic light control system and an inter-university computer network. As the dean of the Faculty of Engineering, he led the establishment of the Research Center for Advanced Science and Technology (RCAST). He held many important positions, such as serving as the president of the Information Processing Society of Japan and the Institute of Electrical Communication Engineers of Japan, the chairperson of the OECD Committee for Scientific and Technological Policy (CSTP), and the first director-general of the National Institute of Informatics.

JABEE (日本技術者教育認定機構, the Japan Accreditation Board for Engineering Education)

A third-party organization that accredits engineering education programs in higher education institutions. It was established in 1999 to join the international mutual recognition of engineering education and became a signatory to the Washington Accord in 2005 (provisionally in 2001). Since its inception, engineering ethics has been an essential requirement in the common criteria. A JABEE program graduate becomes an Engineer-in-Training, designated by the MEXT minister, and is exempt from the first step examination for Gijutsushi.

JFES (日本工学会, the Japan Federation of Engineering Societies)

A Japanese federation of academic societies in all engineering fields. As the first engineering society in Japan, it was founded by Imperial College of Engineering graduates and became a federation of organizational members in 1922. In 1972, it joined WFEO, along with the Science Council of Japan (SCJ), as representatives of Japan and has interacted with engineering societies globally. Based on the awareness of Japan's problems highlighted at the

WFEO meetings, the JFES worked on the international mutual recognition of educational accreditation and established JABEE in 1999, in collaboration with the JSEE.

JSEE (日本工業教育協会, 日本工学教育協会, the Japanese Society for Engineering Education)

An association founded in 1952 to promote engineering education, based on a plan from the American Society for Engineering Education (ASEE) Engineering Education Mission to Japan in 1951. It comprised seven (later eight) regional societies in Japan. The JSEE has promoted collaboration in various engineering education fields to realize industry-academia cooperative education from its inception. From the late 1980s, it worked with the Ministry of Education (MOE) to reform Japan's engineering education and established JABEE in 1999 in collaboration with the JFES. The Japanese name was changed from Kōgyō (工業) to Kōgaku (工学) in 1994.

JUAA (大学基準協会, the Japan University Accreditation Association)

A university association for higher education institutional accreditation, established in 1947 with the support of the Civil Information and Education Section (CI&E) of the General Headquarters, Supreme Commander for the Allied Powers (GHQ/SCAP), modeled on the U.S. accreditation system. Initially, the Japanese university administration adopted the JUAA accreditation standards. However, due to the enactment of the Standards for Establishment of Universities (SEU) in 1956, the MOE, rather than a third-party accreditation, approved Japanese higher education institutions. In the 1990s, due to the increasing demand for university self-examination and self-evaluation, the JUAA started institutional accreditation under a self-study by each university.

KAKENHI (科研費, the Grants-in-Aid for Scientific Research)

A core competitive funding program that promotes all aspects of academic research in Japan, ranging from the humanities and social sciences to the natural sciences. The state budget funds the program.

Keidanren (経団連, the Business Federation)

Japan's largest business organization, founded in 1946. With an organizational membership comprising industrial associations and representative companies of Japan, the federation gathers opinions from the business community to make policy proposals and requests to the government. In 2002, it merged with Nikkeiren to establish a new organization, the Japan Business Federation.

Keizai Doyukai (経済同友会, the Japan Association of Corporate Executives)

A business association for top executives, founded in 1946. All members are individuals. From a business professional perspective, the association has provided progressive policy proposals that are not necessarily confined to respective company interests.

Koga Issac (古賀逸策, 1899–1982)

An electrical engineering professor at the University of Tokyo and Tokyo Institute of Technology. He invented cutting angles to produce a temperature-insensitive quartz oscillation plate. He was deeply involved in the postwar engineering education reform, including the Association for the Advancement of Electrical Engineering Education, the JSEE, and the ASEE Engineering Education Mission to Japan, and devoted attention to engineering ethics.

Kokutai (国体, Japan's national polity)

The national identity of prewar Japan. Based on the idea that the line of the emperors had been unbroken since its foundation, the Japanese nation was regarded as a large family with the emperor and empress as parents. From the end of the Edo period to World War II, *Kokutai* was revered as an absolute concept that proved the uniqueness and excellence of the Japanese nation. In particular, it was emphasized in the 1890 Imperial Rescript on Education; the pedagogical policy had a significant impact on the formation of national morality. After the war, the current Constitution defined the emperor as a symbol of the state; thus, *Kokutai* is no longer used in this sense.

Loyalty and filial piety (忠孝)

Serving one's master faithfully as a subject and taking respectful care of one's parents as a child. These two ideals became a single moral concept based on Confucian thought. They were regarded as a core value of Bushido and the traditional Japanese morality that underpinned the prewar national polity.

Meiji (明治)

The regnal title given to the period of rule of Emperor Meiji (1868–1912).

Mukaibo Takashi (向坊隆, 1917–2002)

A chemical engineering professor at the University of Tokyo. He was appointed the first science attaché at the Embassy of Japan in the United States in 1954 and arranged the U.S.-Japan nuclear cooperation agreement. He contributed internationally to science and technology policy, including the United Nations Advisory Commission on the Application of Science and Technology for Development and the Japan National Commission for UNESCO. Mukaibo also served as the chairperson of the Japan Society for the Promotion of Science (JSPS) Committee 149 for the U.S.-Japan Meetings. He was the dean of the Faculty of Engineering during the student riots in 1968–1969 and later became the university president. After retiring, he became the acting chairperson of the Japan Atomic Energy Commission. He also held several presidential positions at institutions such as the JSEE and the EAJ.

Murakami Yoichiro (村上陽一郎, 1936–)

A professor of science and technology studies at the University of Tokyo and International Christian University. He has published numerous books and translations on the history and philosophy of science, such as *What Is the Scientist?* in 1994. He served as the director of RCAST and has been involved in science and technology policy at various Japanese ministries. Murakami has also served as a member of the OECD/CSTP and the UNESCO Commission on the Ethics of Scientific Knowledge and Technology.

Nikkeiren (日経連, the Japan Federation of Employers' Associations)

An association for business owners, founded in 1948 with the motto: "Employers, Be Righteous and Strong." It aimed to stabilize labor-management relations. In 2002, it merged with Keidanren to establish a new organization, the Japan Business Federation.

Ohashi Hideo (大橋秀雄, 1931–)

A mechanical engineering professor at the University of Tokyo and, later, the president of Kogakuin University. As the vice-chairperson of the JSEE–JFES joint committee established in 1997, he led the establishment of JABEE. He made further efforts to join the Washington Accord and became the second president of JABEE in 2005. In the late 1990s and early 2000s, he also promoted engineering education reform as the director of the SCJ's Section 5, the vice president of the JSEE, and the president of the JFES.

Okamura Sogo (岡村總吾, 1918–2013)

An electrical engineering professor at the University of Tokyo and, later, the president of Tokyo Denki University. He organized Committee 149 as the JSPS president and co-chaired the U.S.-Japan Joint Task Force on Engineering Education with Mildred Dresselhaus of MIT in the 1990s. He also served as the director of the SCJ's Section 5 from 1988 to 1994.

Okochi Masatoshi (大河内正敏, 1878–1952)

A mechanical engineering professor at Tokyo Imperial University and, later, the director of RIKEN. After founding the Industrial Company for RIKEN, he developed the RIKEN Group (RIKEN Konzern) and advocated for a *kagaku shugi kōgyō* (science-based industry) to reform the old-fashioned capitalist industry. He was also a member of the House of Peers and established the Association for the Promotion of Industrial Problems (Kōseikai). He promoted prewar industrial policy in various key positions, which led to his purge from public office after the war.

Onjō shugi (温情主義)

A management-familism that regards the employer-employee relationship as that of a parent and a child and seeks to harmonize the organization based on a paternalistic, hierarchical relationship. "Onjō" means compassion and mercy.

RCAST (先端科学技術研究センター, the Research Center for Advanced Science and Technology)

An affiliated institute of the University of Tokyo, established in 1987. It promotes interdisciplinary research activities based on the following four mottoes: interdisciplinary approach, mobility, international perspective, and openness. Inose Hiroshi, the dean of the Faculty of Engineering, led the planning for the establishment. Murakami Yoichiro served as the director from 1993 to 1995.

Russo-Japanese War (日露戦争, 1904–1905)

The war that broke out between Japan and Russia in 1904 over their respective interests in Manchuria and the Korean Peninsula. Japan won both ground and naval battles, and the Russian Revolution of 1905 broke out, allowing Japan to achieve victory under the Treaty of Portsmouth in 1905. This victory over a Western power gave Japan a great sense of national pride. However, the lack of monetary compensation despite the significant number of casualties and the hefty tax increases caused an explosion of public dissatisfaction and campaigns against the peace treaty occurred across the country.

SCJ (日本学術会議, the Science Council of Japan)

A representative organization of Japanese scholars and scientists in all science fields, as well as the humanities. As part of the postwar democratization policy, the SCJ was established in 1949 as a governmental organization to promote science in governmental and industrial undertakings and the lives of the people by making recommendations to the government, issuing statements, and facilitating academic exchanges locally and internationally. It adopted the Charter for Scientific Researchers in 1980. Engineering was included in the SCJ's Section 5 (now in Section 3, together with the physical sciences), which strived to reform engineering education as a member of WFEO and published a proposal in 1997 recommending the introduction of engineering ethics into engineering education.

SEU's generalization (大学設置基準の大綱化)

The Standards for Establishment of Universities (SEU) stipulate the minimum requirements for research and educational functions and facilities of Japanese universities to grant permission to educational corporations and local government to establish universities. They were enacted in 1956 based on the School Education Law. The SEU's generalization was a drastic relaxation of these standards in 1991 to promote the originality of respective universities while making it compulsory for universities to conduct self-evaluation and publish the results.

Shimizu Kinji (清水勤二, 1898–1964)

An electrical engineering professor at the Nagoya Higher Technical School and, later, the principal of Meiji College of Technology. After World War II, he served as the director-general of the MOE Science Education Bureau and became the president of Nagoya Institute of Technology in 1949. Involved in science and industrial education policies as an MOE official since the prewar period, Shimizu took the initiative in establishing the JSEE and became its president in 1956.

Showa (昭和)

The regnal title given to the period of rule of Emperor Showa (1926–1989).

Taisho (大正)

The regnal title given to the period of rule of Emperor Taisho (1912–1926).

Yoshikawa Hiroyuki (吉川弘之, 1933–)

A professor of precision engineering at the University of Tokyo and later university president from 1993 to 1997. While serving as the JSEE and SCJ president, he founded JABEE as the chairperson of the JSEE–JFES joint committee and became the first JABEE president in 1999. He also held many important positions, such as a member of the Council for Science and Technology, president of the International Council for Science (currently the International Science Council), and president of the National Institute of Advanced Industrial Science and Technology, Japan's largest research institute for the industry.

zaibatsu (財閥)

A major business conglomerate in prewar Japan. Capitalist families, such as Mitsui, Mitsubishi, Sumitomo, Furukawa, and the RIKEN Group (RIKEN Konzern), monopolized capital and promoted Japan's industrialization and economic development. However, after the war, GHQ/SCAP regarded them as constituting an economic basis of prewar militarism and dissolved them in the name of economic democratization.

Bibliography

Adams, Stephen B., and Paul J. Miranti. 2011. "U.S. Expatriates, Postwar Knowledge Transfer and Japanese Telecommunications Revival." *The Role of Expatriates in MNCs Knowledge Mobilization* 27: 131–150.

AIEE. 1951. "Statement of Principles of Professional Conduct of the American Institute of Electrical Engineers." *Electrical Engineer* 70(2): 171–173.

Aizawa Yoshikane, and Taketani Mitsuo. 1969. "Tokyo Daigaku wa 'bunka kakumei' no senjō" [The University of Tokyo is a Battlefield of "Cultural Revolution"]. *Gendai no me* [Modern Perspective] (January): 84–95.

Anan Seiichi. 1987. "Gijutsu rinri kanken" [Opinions on Technology Ethics]. *Technology and Ethics in the Modern World: Nanzan Forum for Social Ethics* 3: 23–35.

———. 1991. "Gijutsu to rinri: on homo utens technologia" [Technology and Ethics: On Human Beings Using Technology]. *Technology and Ethics in the Modern World: Nanzan Forum for Social Ethics* 7: 129–142.

Aoki Seizo. 1969. "Shokunin no kagaku to kagaku no shokunin: daigaku funsō kaiketsu no michi harukanari" [Science of Craftsman and Craftsman of Science: The Way to Solve College Riots Is Far]. *Asahi Journal* 11(17): 17–22.

Aoyama Akira. 1936. "The Civil Engineering in Developing Social Civilization." [In Japanese.] *Journal of the JCES* 22 (2): 1–4.

APEC. 1995a. "Seoul APEC Declaration: Seoul, 14 November 1991." *Selected APEC Documents 1989–1994*. APEC#95-SE-05.1. Singapore: APEC Secretariat. February 1995: 71–74.

———. 1995b. *The Osaka Action Agenda: Implementation of the Bogor Declaration*. APEC SOM Steering Committee on ECOTECH (SCE). December 1995.

———. 2000. *The APEC Engineer Manual: The Identification of Substantial Equivalence*. APEC #00-HR-03.1. Singapore: APEC Engineer Coordinating Committee. November 2000.

Asahara Genshichi. 1966. "Gijutsushi-kai tanjō no koro" [When the JCEA Was Born]. *Consultant* 17: 12–14.

Asahi Shimbun. 1960. "Gakusei ni yobikake: zaikai, 'sangaku-kyōdō ni seijisei" [Call on Students: Politics for 'Industry-Academia Cooperation,' Business World]. July 10, 1960.

ASCE. 1927. *Code of Practice of the American Society of Civil Engineers.* New York: ASCE Headquarters.

Auger, Pierre. 1961. *Tendances actuelles de la recherche scientifique* [Current Trends in Scientific Research]. Paris: UNESCO.

Averch, Harvey. 1981. "A. Beikoku kagaku seisaku no kihon genri (Kagaku seisaku no nichibei hikaku (I): II. Kagaku seisaku no wakugumi)" [A. Basic Principles of Science Policy in the U.S. (U.S.-Japan Comparison of Science Policy (I): II. The Framework of the Science Policy)]. *Gakujutsu geppō* [Japanese Scientific Monthly] 34(3): 182–185.

Banno Junji. 2005. *Meiji Democracy.* [In Japanese.] Tokyo: Iwanami Shoten.

Baum, Robert J. 1980. *Ethics and Engineering Curricula.* Hastings-on-Hudson, NY: Hastings Center.

Cabinet Secretary. 1953. "Gijutsushi-hō yōkō ni taisuru iken" [Comments on the Summary of the Japanese CE Act]. November 9, 1953. IPEJ, Tokyo.

Carroll, William. 1995. "Resolution on Accreditation: President, WFEO to All Members of WFEO." April 18, 1995. JSEE, Tokyo.

CCS/R&D. 1949. "The Need for a Management Training Course in the Communications Manufacturing Industry, 6 August 1949." Box 6, folder 16, Kenneth Hopper Papers on Management, Drucker Institute. http://ccdl.libraries.claremont.edu/cdm/search/collection/khp/searchterm/khp00916.pdf.

Central Council for Education. 1957. *Kagaku gijutsu kyōiku no shinkō hōshin ni tsuite* [On the Promotion Measures for Science and Technology Education]. Tokyo: MOE.

———. 1963. *Daigaku kyōiku no kaizen ni tsuite* [On the Improvement of University Education]. Tokyo: MOE.

Chargaff, Erwin. 1980. *Heraclitean Fire: Sketches from a Life before Nature.* Translated by Murakami Yoichiro. Tokyo: Iwanami Shoten.

Chemical & Engineering News. 1948. "Engineers' Council Offers Canons of Ethics." 26(31): 2259.

Chiku Kakugyo. 1983a. *The Dawn of Science and Technology: Books that Changed the World.* [In Japanese.] Nonoichi: KIT.

———. 1983b. "Kikōbon ni miru kagaku gijutsushi" [History of Science and Technology in the Rare Books]. *Back Up* 3: 83–88.

Chittenden, Russell H. 1928. *History of the Sheffield Scientific School of Yale University 1846–1922.* New Haven: Yale University Press.

Council for Industrial Rationalization. 1951. *Waga kuni sangyō no gōrika ni tsuite* [On the Rationalization of Our Nation's Industry]. Tokyo: Trade Enterprise Bureau of the MITI.

———. 1956. *Sangaku-kyōdō kyōiku seido ni tsuite* [On the Industry-Academia Cooperation Education System]. Tokyo: MITI.

CST. 1960. *Jūnengo o mokuhyō to suru kagaku gijutsu sinkō no sōgōteki kihon hōsaku ni tsuite* [On the Comprehensive Basic Measures to Promote Science and Technology Targeting Ten Years Later]. Tokyo: CST.

Daini Rincho. 1982. *Rinchō kihon teigen: Rinji Gyōsei Chōsakai dai 3 ji tōshin* [Basic Proposal: The Third Report of Rincho]. Tokyo: Institute of Administrative Management.

Davis, Michael. 1998. *Thinking Like an Engineer: Studies in the Ethics of a Profession*. New York: Oxford University Press.

Dees, Bowen C. 1997. *The Allied Occupation and Japan's Economic Miracle: Building the Foundations of Japanese Science and Technology 1945–52*. Surrey: Japan Library.

Downey, Gary L., Juan C. Lucena, and Carl Mitcham. 2007. "Engineering Ethics and Identity: Emerging Initiatives in Comparative Perspective." *Science and Engineering Ethics* 13: 463–487.

EAJ. 1996. "Engineering Ethics Education in the U.S.: Aiming for Professional Autonomy" (Heinz C. Luegenbiehl's Lecture, 82nd "Danwa Salon" meeting of the EAJ, July 17, 1996). *EAJ Information* 61.

———. 2000. "Engineering Ethics in Japanese Corporations: A View from the Field" (Scott Clark's lecture, 115th "Danwa Salon" meeting of the EAJ, April 14, 2000). *EAJ Information* 97.

EAJ Engineering Education Committee. 1990. *Asu o sasaeru jinzai ikusei to taisei seibi* [Development of Human Resources and Systems to Support Tomorrow]. Tokyo: EAJ.

Economic Council. 1999. "Keizai shakai no arubeki sugata to keizai shinsei no seisaku hōshin" [Ideal Socio-economy and Policies for Economic Rebirth]. Tokyo: Cabinet Office. November 1999. https://www5.cao.go.jp/99/e/19990705e-keishin.html.

Economic Planning Agency. 1999. *Keizai shakai no arubeki sugata to keizai shinsei no seisaku hōshin* [The Ideal State of the Economic Society and the Policy for the New Economy]. Tokyo: National Printing Bureau.

Editorial Committee for the Unofficial Communications History, ed. 1962. *Teishin shiwa* [Historical Stories of Communications]. Tokyo: Telecommunications Association.

Engineering College Students' Union. 1960. "Zadankai: gakusei to sangaku-kyōdō" [Round-Table Discussion: Students and Industry-Academia Cooperation]. *Japan Science and Technology* 7 (December): 30–33.

Faculty of Engineering, University of Tokyo. 1970. *Kōgakubu news* [Faculty of Engineering News]. April 1, 1970.

Fiske, John. 1894. *Edward Livingston Youmans: Interpreter of Science for the People*. New York: D. Appleton.

FPSIRJ. 1932. Drafts of the Charter, Plans, and Instructions of the Establishment of Gakujutsu Sangyō Shinkōin. [In Japanese.] May 1932. Gakushin-San-02. JSPS, Tokyo.

Fudano Jun. 1995. *Final Research Report of "A Transdisciplinary Study on Public Acceptance of Advanced Science and Technology in Modern Society."* [In Japanese.] Grants-in-Aid for Scientific Research on Priority Areas in FY1991–1994, 06201205.

———. 1996. "Beikoku ni okeru kōgaku rinri kyōiku" [Engineering Ethics Education in the United States]. *Hinshitsu kanri* [Total Quality Management] 47(10): 10–16.

———. 1999. *Final Research Report of "A Study on Professional Ethics Education in the University Curriculum for Engineering and Science."* [In Japanese.] Grants-in-Aid for Scientific Research (C) in FY1994–1995, 06680186.

———, ed. 2015. *Atarashii jidai no gijutsusha rinri* [Engineering Ethics in the New Era]. Tokyo: Foundation for the Promotion of the Open University of Japan.

Fujie Kunio. 1994a. "Conservation of Conception from Technology to Gido: For Conservation of Beautiful Environment." [In Japanese.] *Journal of the JSME* 97(903): 40.

———. 1994b. "Ethics of Technology: Toward the Global Sustainable Development." [In Japanese.] *Journal of the JSME* 97(913): 16.

———. 1996. "Ethics of Engineers: Switch of Conception from Technology to Gido." [In Japanese.] *JSME Proceedings IV* 96(15): 433–434.

———. 2004. *Jitsugaku no rinen to kigyō no susume: Fukuzawa Yukichi to kagaku gijutsu* [Spirit of Practical Study and Entrepreneurship: Fukuzawa Yukichi and Science and Technology]. Tokyo: Keio University Press.

Fujiki Atsushi. 2011. "External Causes of World-Wide Spread of Engineering Ethics: From the Viewpoint of Internationalization of Engineer Qualification and Accreditation System for Engineering Education." [In Japanese.] *Journal of Innovative Ethics* 4: 50–66.

———. 2012. "How Should We Accept the American-Style Engineering Ethics?" *Proceedings of the Second International Conference on Applied Ethics and Applied Philosophy in East Asia*, 47–60. Dalian: Dalian University of Technology Press. http://www.kurume-nct.ac.jp/GEM/individual/fujiki/files/Fujiki2011.pdf.

Fukuzawa Yukichi. (1893) 1961a. "Gishi shakai" [Engineers' Society]. In Vol. 14 of *Fukuzawa Yukichi zenshū* [Complete Works of Fukuzawa Yukichi], 184–186. Tokyo: Iwanami Shoten.

———. (1895a) 1961b. "Gishi no tokugi" [Virtue of Engineers]. In Vol. 15 of *Fukuzawa Yukichi zenshū* [Complete Works of Fukuzawa Yukichi], 324–326. Tokyo: Iwanami Shoten.

———. (1895b) 1961c. "Gishi no shin'yō" [Trust of Engineers]. In Vol. 15 of *Fukuzawa Yukichi zenshū* [Complete Works of Fukuzawa Yukichi], 328–329. Tokyo: Iwanami Shoten.

Furuki Moriyasu, and Sakamoto Shinji. 2004. "Doboku Gakkai rinri kitei to gijutsusha undō" [The JSCE's Code of Ethics and the Engineers' Movement]. *Journal of the JSCE* 89(5): 71–73.

Furuya Keiichi. 1971. "Daigaku towa nanika" [What Is University?]. In Moriguchi 1971: 13–39.

———. 2003. "America no gakkai rinri kōryō no genten to sono tenkai" [The Origin and Development of the Code of Ethics of Academic Societies in America]. *Journal of Japan Society for Atmospheric Environment* 38(1): A4–A14.

Fushimi Koji. 1980. "Kagakusha shakai no kenshō" [The Charter of the Society of Scientists]. In Watanabe and Igasaki 1980: 11–21.

GHQ/SCAP. 1990. "Communication." In Vol. 55 of *History of the Nonmilitary Activities of the Occupation of Japan, 1945–1951*. Tokyo: Nihontosho Center.

Gijutsu Senmonka Katsudo Shisatsudan [Engineering Professionals' Activity Survey Team]. 1963. *Activities of P.E. and C.E. in U.S.A.* [In Japanese.] Tokyo: Gijutsu Senmonka Katsudo Shisatsudan.

Gijutsushi Council. 2000. *Gijutsushi seido no kaizen hōsaku ni tsuite* [On Measures to Improve the Gijutsushi System]. Tokyo: STA.

Godin, Benoît. 2017. *Models of Innovation: The History of an Idea.* Cambridge, MA: MIT Press.

Graduate Students Union of Urban Engineering of the University of Tokyo. 1969. "Jiritsu teki kenkyūsha o mezashite no tatakai: Tōdai tōsō ni okeru toshikōgaku daigaku insei no shuchō" [Struggle for Independent Researcher: Insistence of Graduate Students at Urban Engineering in the Todai Struggle]. *Shizen* [Nature] 24(4): 52–61.

Gunn, Alastair S., and P. Aarne Vesilind. 1986. *Environmental Ethics for Engineers.* Boca Raton: CRC Press.

Harada Kosaku. 1991. "Beikoku ABET no engineering accreditation ni tsuite" [About the Engineering Accreditation of ABET]. *Journal of the JSEE* 39(4): 44–47.

———. 1996. "Sekai kakkoku no kōgaku kyōiku no hyōka nintei to senmon gijutsusha shikaku no kankei: global engineer shikaku eno michi" [Relationship Between Accreditation of Engineering Education and Professional Engineers' Qualifications in the World: The Road to Global Engineers' Qualification]. *Journal of the JSEE* 44(1): 45–49.

Harding, Francis C., and Donald T. Canfield. 1936. *Legal and Ethical Phases of Engineering.* New York: McGraw-Hill.

Harris, Charles E., Jr., Michael S. Pritchard, and Michael J. Rabins. 1995. *Engineering Ethics: Concepts and Cases.* Belmont, CA: Wadsworth.

———. 1998. *Kagaku gijutsusha no rinri: sono kangaekata to jirei* [Ethics for Science-Engineers: Concepts and Cases]. Translated and edited by the JCEA. Tokyo: Maruzen.

———. 2000. *Kagaku gijutsusha no rinri: sono kangaekata to jirei* [Ethics for Science-Engineers: Concepts and Cases], 2nd ed. Translated and edited by the JCEA. Tokyo: Maruzen.

———. 2009. *Engineering Ethics: Concepts and Cases*, 4th ed. Belmont, CA: Wadsworth.

Hashimoto Takehiko. 1999. "The Hesitant Relationship Reconsidered: University-Industry Cooperation in Postwar Japan." In *Industrializing Knowledge: University-Industry Linkages in Japan and the United States*, edited by Lewis M. Branscomb, Kodama Fumio, and Richard Florida, 234–251. Cambridge, MA: MIT Press.

Hattori Hiroyasu. 1996. "Gijutsu Hon'yaku Center" [Report of the TTC]. *Gijutsushi: Consulting Engineer* 339 (May): 25.

———. 1997a. "Gijutsu Hon'yaku Center" [Report of the TTC]. *Gijutsushi: Consulting Engineer* 359 (October): 32.

———. 1997b. "Gijutsu Hon'yaku Center" [Report of the TTC]. *Gijutsushi: Consulting Engineer* 360 (November): 30.

———. 1998. "Gijutsu Hon'yaku Center" [Report of the TTC]. *Gijutsushi: Consulting Engineer* 367 (June): 40.

Hazama Hiroshi. 1963. *Nihonteki keiei no keifu* [Genealogy of Japanese Management]. Tokyo: Japan Management Association.

———. 1978. *Nihon ni okeru rōshi kyōchō no teiryū* [The Bottom Flow of the Labor-Management Harmonization in Japan]. Tokyo: Waseda University Press.

Hirayama Fukujiro. 1926. "Jitsubutsu kyōiku, gainen kyōiku, ningen kyōiku no ketsujo" [Lack of Real Education, Concept Education, and Human Education]. In *Kōgyō kyōiku no kenkyū* [Research on Engineering Education], edited by Koseikai, 26. Tokyo: Koseikai Shuppanbu.

———. 1960. "Gijutsushi-hō no mondaiten ni tsuite" [On the Problems of the Japanese CE Act]. *Mechanization of Construction* 130: 2–4.

———. 1961. "Consulting engineer towa nanika" [What Is the Consulting Engineer?]. *Living in Civil Construction*, 133–151. Tokyo: Sankaido.

Hirooka Masaaki. 1990. "Ōbei ni okeru kōgaku kyōiku kaikaku no genjō: Nihon Kōgaku Academy no kaigai chōsa kara" [Current Situation of Engineering Education Reform in Europe and the United States: From the EAJ's Overseas Survey]. *Daigaku to gakusei* [University and Student] 291: 20–27.

Hirose Hitoshi. 2011. "JSPE hossoku made no 105 nichikan" [105-Day Road up to the JSPE Establishment]. In JSPE 2011: 17–19.

Hiroshige Tetsu. 1961. "Daigaku jin to sangyō jin" [University People and Industry People]. *Chuokoron* 76(5): 100–110.

———. 1965. "Kagaku gijutsu kihonhō wa hitsuyō ka" [Is the Basic Act on Science and Technology Necessary?]. *Kagaku Asahi: A Monthly Journal of Science* 25(12): 134–135.

Hiyagon Hitoshi. 2015. "Nihon no gijutsusha seido henkaku teitai to konran: sono mondai bunseki to kaiketsusaku no teiji" [The Stagnation and Confusion in Japan's Engineering System Reform: Its Analysis and Solutions]. PhD Dissertation, Nagoya University.

Hojo Tokiyuki. 1913. "Boy Scout undō ni tsukite" [About the Boy Scout Movement]. *Kyōiku kenkyūkai kōenshū* [Proceedings of the Educational Research Group] 6, edited by the Educational Research Group of the Hiroshima Higher Normal School, 18–33.

Honda Naoshi. 1989. *Gijutsushi eno izanai* [Invitation to Gijutsushi]. Tokyo: Techno-Communications.

Hopper, Kenneth. 1982. "Creating Japan's New Industrial Management: The Americans as Teachers." *Human Resource Management* 21(2–3): 13–34.

Hopper, Kenneth, and William Hopper. 2007. *The Puritan Gift: Reclaiming the American Dream amidst Global Financial Chaos*. New York: I.B. Tauris.

Hori Yasuaki. 1994a. "Gijutsushi no identity to Shōgai Iinkai" [The Gijutsushi's Identity and the External Affairs Committee]. *Gijutsushi: Consulting Engineer* 315 (June): 59–60.

———. 1994b. "Gairan ni taisuru sekkyokuteki ōtai ni tsuite" [Proactive Measure against the Disturbance]. *Gijutsushi: Consulting Engineer* 318 (September): 17.

Ichikawa Atsunobu, et al. 1990. "Atarashii kōgaku kyōiku no kakuritsu o mezasite" [Toward the Establishment of a New Engineering Education]. *Daigaku to gakusei* [University and Student] 291: 4–19.

IEICE/FACE. 1997. "Denshi Jōhō Tsūshin Gakkai rinri kōryō shian: sono kaisetsu to sakutei no keii" [Proposal of the IEICE Code of Ethics: The Explanation and the Process of Formulation]. *IEICE Technical Report* 97(84): 26–44.

———. 1999. "Denshi Jōhō Tsūshin Gakkai rinri kōryō kaisetsu" [Explanation of the IEICE Code of Ethics]. *Journal of the IEICE* 82(2): 161–174.

Igasaki Akio. 1980. "Kagakusha Kenshō to Kagaku Gijutsu Kihonhō" [The Charter for Scientific Researchers and the Basic Act on Scientific Research]. In Watanabe and Igasaki 1980: 71–104.

Ikegami Yoshihiko. 1996. "Henshū kōki" [Editorial Postscript]. *Gendai shisō* [Revue de la pensée d'aujourd'hui] 24(6): 342.

Ikeuchi Satoru. 1996. "Yakusha atogaki" [Afterword by Translator]. In NAS, NAE, and the Institute of Medicine. 1995. *On Being a Scientist: Responsible Conduct in Research*, 2nd ed. Translated and edited by Ikeuchi Satoru, 87–90. Kyoto: Kagakudojin.

Imai Kaneichiro. 1980. "1980 nendai, kyōryoku no jidai" [The 1980s, the Age of Cooperation]. *Machinist* 24(1): 14–15.

———. 1994. "Accreditation in Engineering in Japan is on the Way." *Ideas* 2 (December): 24–29.

———. 1995. "Kōgaku kyōiku kokusaika no mondaiten" [Problems of Engineering Education's Internationalization]. *Journal of the JSEE* 43(4): 1–3.

———. 2011. "JSPE no 10 nen ni omou" [In Honor of the Tenth Anniversary of the JSPE]. In JSPE 2011: 56–59.

Inagaki Eizo, et al. 1971. "Kōgakubu kaikaku eno teigen" [Proposal for the Reformation of the Faculty of Engineering]. In Moriguchi 1971: 141–205.

Industry-Academia Cooperation Committee. 1959. "Sangaku-Kyōdō Iinkai hossoku" [The Industry-Academia Cooperation Committee Was Established]. *Journal of the JSEE* 6(2): 14–20.

Inose Hiroshi. 1986a. "I. Nichibei kaigi no gaiyō" [I. Overview of the U.S.-Japan Meeting]. *Gakujutsu geppō* [Japanese Scientific Monthly] 39(1): 6–9.

———. 1986b. "Sentan gijutsu to kokusai kankyō" [Advanced Technology and the International Environment]. *Joho kanri: Journal of Information Processing and Management* 29(5): 363–377.

———. 1987a. "I. Kaigi no haikei to komyunike no gaiyō" [I. Background of the Meeting and Overview of the Communique]. *Gakujutsu geppō* [Japanese Scientific Monthly] 40(2): 84–88.

———. 1987b. *Jōhō no seiki o ikite* [Living the Century of Information]. Tokyo: University of Tokyo Press.

———. 1990. *Bunka to shiteno kagaku gijutsu o kangaeru* [Thinking About Science and Technology as a Culture]. Tokyo: Mita Shuppankai.

———. 1994. "High-tech keizai o sasaeru kiso kenkyū no shinkō ni oite kongo 25 nen ni daigaku ga hatasu yakuwari: nihon no tenbō" [The Role of Universities in the Promotion of Basic Research Supporting the High-Tech Economy in the Next Twenty-Five Years: Japan's Perspective]. *Gakujutsu geppō* [Japanese Scientific Monthly] 47(7): 15–22.

Inose Hiroshi, and Murakami Yoichiro. 1992. *Kenkyū kyōiku system* [Research and Education System]. Tokyo: Asakura Publishing.
Inose Hiroshi, Nishikawa Tetsuji, and Uenohara Michiyuki. 1982. "Cooperation Between Universities and Industries in Basic and Applied Science." In *Science Policy Perspectives: USA-Japan*, edited by Arthur Gerstenfeld, 43–61. New York: Academic Press.
Investigation Collaborator's Council on the Promotion of Engineering Education. 1989. *Henkaku ki no kōgaku kyōiku* [Engineering Education in the Transitional Phase]. Tokyo: MOE.
IPEJ. 2001a. *Kagaku gijutsu ni kakawaru moral ni kansuru chōsa kenkyū hōkokusho* [Survey Report on Morals Related to Science and Technology]. Tokyo: IPEJ.
———. 2001b. *Nihon Gijutsushi-kai 50 shūnen kinenshi* [Fiftieth Anniversary of the IPEJ]. Tokyo: IPEJ.
IPSJ. 1997. *Rinri Kōryō Chōsa Iinkai hōkokusho* [Report of the Code of Ethics Investigation Committee]. Tokyo: IPSJ.
JABEE. 1999. *"Nihon gijutsusha kyōiku nintei seido no kakuritsu o mezashite: kokusaiteki ni tsūyōsuru gijutsusha kyōiku no tameni"* [Aiming for the Establishment of the Japanese Engineering Education Accreditation System: For Engineering Education with International Consistency]. Tokyo: JSEE.
———. 2000. "Nihon gijutsusha kyōiku nintei kijun shikōyō V 1.0" [Japanese Accreditation Criteria for Engineering Education, Trial Version 1.0]. *Journal of the JSEE* 48(1): 12–20.
———. 2003. *Nihon Gijutsusha Kyōiku Nintei Kikō (JABEE) wa ikanishite umaretaka* [How JABEE Was Born]. Tokyo: JABEE.
———. 2012. *JABEE no ayumi: setsuritsu kara 13 nen (1999–2012)* [History of JABEE: Thirteen Years from the Establishment (1999–2012)]. Tokyo: JABEE.
Japan Chemical Daily. 1993. "Engineer no pro shikaku seido o dōnyū" [Introducing Professional License System for Engineers]. January 19, 1993.
Japan PE Council. 1994. *Kokusai shikaku, Professional Engineer eno michi* [Road to the Professional Engineer, an International License]. Tokyo: Diamond.
Japanese Central Executive Committee. 1951. *Report of the Institute for Engineering Education*. Tokyo: Japanese Central Executive Committee.
Japanese Society for Ethics. 1985. *Gijutsu to rinri* [Technology and Ethics]. Tokyo: Ibunsha.
JCEA. 1951a. "Gijutsushi-kai tanjō" [The Birth of the JCEA]. *Gijutsu Service* [Engineering Service] 1(1): 1.
———. 1951b. "Gijutsushi Fukumu Yōkō" [The Code of Service for CEs]. *Gijutsu Service* [Engineering Service] 1(2): 8.
———. 1955. "(Shiryō) Kokusai Gijutsushi Renmei (FIDIC) no gaiyō" [(Ref.) Summary of FIDIC]. *JCEA* 3: 6–9.
———. 1960a. Minute of the Second Regular General Meeting. [In Japanese.] *JCEA* 12: 4–7.
———. 1960b. Minute of the Seventh Board Meeting in FY1960. [In Japanese.] October 11, 1960. IPEJ, Tokyo.

———. 1960c. Minute of the Ninth Board Meeting in FY1960. [In Japanese.] November 9, 1960. IPEJ, Tokyo.
———. 1960d. "Nihon Gijutsushi-kai shusai Ōshū chōsadan hōkoku" [Report of the JCEA's Survey Team on Europe]. *JCEA* 17.
———. 1961a. Minute of the Eleventh Board Meeting in FY1960. [In Japanese.] January 10, 1961. IPEJ, Tokyo.
———. 1961b. Minute of the Twelfth Board Meeting in FY1960. [In Japanese.] February 14, 1961. IPEJ, Tokyo.
———. 1961c. Minute of the Thirteenth Board Meeting in FY1960. [In Japanese.] March 14, 1961. IPEJ, Tokyo.
———. 1961d. "(Shiryo) Oranda Gijutsushi Kyōkai (O.N.R.I.) kara no chōsa jikō ni taisuru kaitō" [(Ref.) Reply from the Netherlands Consulting Engineers Association (ONRI) to the Survey Items]. *JCEA* 22: 11–17.
———. 1970. Minute of the First Board Meeting in FY1970. [In Japanese.] April 14, 1970. IPEJ, Tokyo.
———. 1981. *Nihon Gijutsushi-kai sanjū nenshi* [Thirty-Year History of the JCEA]. Tokyo: JCEA.
———. 1992a. "Iinkai hōkoku: Shōgai Iinkai" [Report: the External Affairs Committee]. *Gijutsushi: Consulting Engineer* 287 (May): 24.
———. 1992b. "Gijutsusha shikaku no kokusai seigōsei ni taishite Nihon Gijutsushi-kai ga shutaisei o motsutameno sochi ni tsuite" [On Measures for the JCEA to Take Initiative in International Consistency of Engineer Licenses]. *Gijutsushi: Consulting Engineer* 294 (November): 22–23.
———. 1994. *Gijutsusha no yōsei, kakuho ni kansuru chōsa II* [Survey on Training and Securing of Engineers II]. Tokyo: JCEA.
———. 1998. "Shaji" [Acknowledgements]. In Harris, Jr., et al. 1998: 482.
———. 2002. "Yakusha atogaki" [Afterword by Translators]. In Harris, Jr., et al. 2000: 423–429.
JCEA External Affairs Committee. 1993. "Beikoku PE seido dōnyū mondai ni tsuite no Shōgai Iinkai no taisho" [External Affairs Committee's Addressing the Issue of Introducing the American PE System (Interim Report)]. *Gijutsushi: Consulting Engineer* 305 (September): 21–23.
———. 1994a. "Sōgō hōkoku: Shōgai Iinkai no setsuritsu to keika" [General Report: Establishment and Progress of the External Affairs Committee]. *Gijutsushi: Consulting Engineer* 315 (June): 60–61.
———. 1994b. "Nihon Kōgyō Gijutsu Shinkō kyōkai setsuritsu no Nihon PE Kyōgikai ni kansuru kenkai" [Opinions on the Japan PE Council established by the JTTAS]. *Gijutsushi: Consulting Engineer* 316 (July): 16.
———. 1994c. "Beikoku PE Shiken ni kansuru genjō hōkoku" [Status Report on the U.S. PE Examination]. *Gijutsushi: Consulting Engineer* 320 (November): 22.
JCEA/TTC. 1998. Agenda, 5 October 1998, Private Collection of Fukumoto Muneki.
JCES. 1933a. "Proceedings of the Society." [In Japanese.] *Journal of the JCES* 19(2): 1–14.
———. 1933b. "Proceedings of the Society." [In Japanese.] *Journal of the JCES* 19(4): 1–22.

———. 1936a. "Proceedings of the Society." [In Japanese.] *Journal of the JCES* 22(6–7): 1–12.

———. 1936b. "Proceedings of the Society." [In Japanese.] *Journal of the JCES* 22(12): 1–13.

———. 1938a. "Proceedings of the Society." [In Japanese.] *Journal of the JCES* 24(2): 1–17.

———. 1938b. "Doboku hōkoku undō" [Repay Movement for the Nation with Civil Engineering]. *Journal of the JCES* 24(2): 1.

———. 1938c. "Announcement." [In Japanese.] *Journal of the JCES* 24(3): 1–12.

JEA. 1942. "Nihon Nōritsu Kyōkai Setsuritsu" [The Establishment of the JEA]. *Nihon nōritsu* [Japan Efficiency] 1(1): 65.

JFES. 1992. Agenda for the Regular General Meeting in FY 1992. April 22, 1992. JFES, Tokyo.

———. 1995. "Gijutsusha no Shakaiteki Hyōka Kōjō ni tsuite no Tokubetsu Iinkai hōkoku (chūkan trimatome)" [Interim Report of the Special Committee on Increasing Engineers' Esteem]. October 1995. JSEE, Tokyo.

JMF and ENAA. 1991. *Kakkoku ni okeru engineer shikaku no jittai chōsa: engineering nōryoku no kyōka ni kansuru chōsa kenkyū hōkokusho* [Survey of Engineering Qualifications: Research Report on Strengthening Engineering Abilities]. Tokyo: JMF and ENAA.

Johnson, Chalmers. 1982. *MITI and the Japanese Miracle: The Growth of Industrial Policy, 1925–1975*. Stanford, CA: Stanford University Press.

JPC. 1958. *America sangaku-kyōdō no jittai: Sangaku-Kyōdō Senmon Shisatsudan hōkokusho* [Actual Situation of American Industry-Academia Cooperation: A Report of Industry-Academia Cooperation Expert Visiting Team]. Tokyo: JPC.

———. 1960. "Gutaika no michi isogu sangaku-kyōdō: kakukai no ugoki to enquête chōsa ni miru" [Industry-Academia Cooperation to Hurry the Way to the Implementation: Examining the Movements of Each Field and the Questionnaire Survey]. *Productivity* 152 (February): 20–22.

JSCE. 1994. *Doboku Gakkai no 80 nen* [Eighty Years of the JSCE]. Tokyo: JSCE.

JSEE. 1953. Agenda of Board Meetings and Steering Committee Meetings. [In Japanese.] *Journal of the JSEE* 1(1): 87–104.

———. 1956a. Eighteenth Steering Committee Meeting. [In Japanese.] *Journal of the JSEE* 3(2): 147–151.

———. 1956b. Board Meeting and Board of Representatives in FY1955. [In Japanese.] *Journal of the JSEE* 4(1–2): 227–237.

———. 1965. "Panel tōron: sangaku-kyōdō o kakuritsu shiyō" [Panel Discussion: For Establishing Industry-Academia Cooperation]. *Journal of the JSEE* 12(1–2): 37–72.

———. 1969. "Kaihō" [Newsletter: The Minutes of the Committee of Educational Policy]. *Journal of the JSEE* 16(2): 86–95.

———. 1970. "Panel tōgi: kōgakukei daigaku ni okeru sangaku-kyōdō no hitsuyō to sono gutaisaku" [Panel Discussion: The Necessity and Concrete Measures of Industry-Academia Cooperation in Engineering Colleges]. *Journal of the JSEE* 18(1): 23–43.

JSF. 1989. *30 nen no ayumi* [Thirty Years of History]. Tokyo: JSF.
JSPE. 2011. *JSPE 10 shūnen kinenshi: saisho no 10 nen* [Tenth Anniversary of the JSPE: The First Decade 2000–2010]. Tokyo: JSPE.
JSPS. 1998. *Nihon Gakujutsu Shinkōkai 30 nenshi* [Thirty-Year History of the JSPS]. Tokyo: JSPS.
JSPS Committee 149. 2000. *Sentan Gijutsu to Kokusai Kankyō Dai 149 Iinkai: katsudō no rekishi 1984–2000* [Committee 149 on Advanced Technology and the International Environment: History of Activities 1984–2000]. Tokyo: JSPS.
JSPSR. 1939. *Annual Report of the JSPSR for 1937*. [In Japanese.] Tokyo: JSPSR.
———. 1941. *Annual Report of the JSPSR for 1940*. [In Japanese.] Tokyo: JSPSR.
JUSE. 1996. *Hinshitsu kanri* [Total Quality Management]. 47(10).
Kagawa Chiaki, and Komatsu Yoshihiko, eds. 2014. *Seimei rinri no genryū: sengo nihon shakai to bioethics* [The Origin of Bioethics: Postwar Japanese Society and Bioethics]. Tokyo: Iwanami Shoten.
Kajino Shinichi. 1996. "Daigaku kaikaku to kōgaku kyōiku" [University Reform and Engineering Education]. *Journal of the JSEE* 44(5): 15–23.
Kakihara Yasushi. 1996. "Science versus Practice in Engineering Education of Modern Japan: Telegraphy at the Imperial College of Engineering, Tokyo." [In Japanese.] *Japan Journal for Science, Technology & Society* 5: 1–20.
Kanpo Official Gazette. 2000. "On the Amendment of the PE Act." [In Japanese.] extra ed. 83 (April 26): 6–7.
Karaki Junzo. 1980. *"Kagakusha no shakai teki sekinin" ni tsuite no oboegaki* [Note on "the Social Responsibility of Scientists"]. Tokyo: Chikuma Shobo.
Kasahara Masao. 1995. "Review of Ethics on Advanced Communications." [In Japanese.] *IEICE Technical Report* 95(64): 33–40.
Kase Kosaku. 1953. "Shokugyōjin toshite daigaku e kibō suru" [My Hope for University, as a Worker]. *Journal of the JSEE* 1(2): 56–61.
Kato Ichiro. 1969. *"Nana gakubu daihyōdan tono kakuninsho" no kaisetsu* [Guide to "Confirmation with the Delegates of Seven Faculties"]. Tokyo: University of Tokyo Press.
Kawano Yasuo. 1981. "Nihon Gijutsushi-kai to Hirayama Fukujiro-san" [Mr. Hirayama Fukujiro and the JCEA]. In JCEA 1981: 14.
Kawasaki Shoichiro. 1969. "Sangaku-kyōdō gungaku-kyōdō to gakumon kenkyū no jiyū" [Industry-Academia Cooperation, Military-University Cooperation and Academic Research Freedom]. *Cultural Review* 90 (March): 67–80.
Keidanren. 1996a. *An Attractive Japan: Keidanren's Vision for 2020*. [In Japanese.] Tokyo: Keidanren. https://www.keidanren.or.jp/japanese/policy/vision/.
———. 1996b. "Developing Japan's Creative Human Resources: An Action Agenda for Reform in Education and Corporate Conduct." [In Japanese.] Tokyo: Keidanren. https://www.keidanren.or.jp/japanese/policy/pol083/.
Keizai Doyukai. 1969. *Kōji fukushi shakai no tame no kōtō kyōiku seido* [Higher Educational System for Higher Welfare Society]. Tokyo: Japan Association of Corporate Executives.
———. 1976. *Keizai Dōyūkai sanjū nenshi* [Thirty-Year History of Keizai Doyukai]. Tokyo: Japan Association of Corporate Executives.

———. 1997. *Manifest for a Market-Oriented Economy: Action Program for Japan Toward the 21st Century.* Tokyo: Keizai Doyukai. January 1997. https://www.doyukai.or.jp/en/policyproposals/1996/pdf/970109a.pdf.

Kensetsu Tsushin Shimbun. 1997. "Gijutsusha sōgo ninshō APEC Engineer sōsetsu" [APEC Engineer, a Mutual Recognition for Engineers, Was Established]. September 2, 1997.

Kihara Junji. 1969. "Daigaku kaikaku no hōkō" [Direction of the University Reformation]. In Moriguchi 1969: 348–385.

Kikuchi Shigeaki. 2004. "Gijutsusha rinri no rekishiteki haikei" [The Historical Background of Engineering Ethics]. *Journal of Japanese Scientist* 39(1): 4–9.

Kimura Hisao. 1976. "America Denki Gakkai (IEEE) no rinri no hōten" [The Code of Ethics of American Society of Electricity (IEEE)]. *Journal of the IEEJ* 96(9): 769–770.

———. 1977a. "'Rinri no hōten' ni kansuru shokan" [Comments on the "Code of Ethics"]. *Journal of the JSEE* 25(1): 35–40.

———. 1977b. "'Rinri no hōten' ni kansuru shokan" [Comments on the "Code of Ethics"]. *Journal of the IEEJ* 97(4): 249–250.

———. 1978. "Gijutsusha rinri to shuken zaimin" [Engineering Ethics and Popular Sovereignty]. *Electrical Review* [Denki hyōron] 63(9): 52–53.

———. 1979. "Gijutsusha rinri to shuken zaimin no kankei" [Relationship Between Engineering Ethics and Popular Sovereignty]. *Journal of the JSEE* 27(1): 43–46.

Kimura Katsusaburo. 1992. "Iinkai hōkoku: Shōgai Iinkai" [Report: the External Affairs Committee]. *Gijutsushi: Consulting Engineer* 295 (December): 22.

———. 2007. "Shōgai Iinkai deno katsuyaku" [Remarkable Activities in the External Affairs Committee]. In Umeda 2007: 110–111.

KIT. 1980. *Kanazawa Kōgyō Daigaku: gakusei binran* [Student Handbook of KIT]. Nonoichi: KIT.

———. 1995. *Core Guide Book 1995.* [In Japanese.] Nonoichi: KIT.

Kitami Akihiko. 1962. "Waga kuni ni okeru sangaku-kyōdō no sokushin jōkyō" [Current Situation of the Promotion of Industry-Academia Cooperation in Japan]. *Japan Science and Technology* 26 (July): 9–14.

Ko-Yamakawa Danshaku Kinenkai [Memorial Committee of Late Baron Yamakawa], ed. 1937. *Danshaku Yamakawa sensei ikō* [Papers of Baron Yamakawa]. Kyoto: Ko-Yamakawa Danshaku Kinenkai.

Kobayashi Hiroomi. 2005. "Introductory Article for Professional Engineers' Translation Center." [In Japanese.] *IPEJ Journal* 17(8): 28–29.

Kobayashi Koji. 1987. "Nihon Kōgaku Academy no hossoku ni atatte" [On the Establishment of the EAJ]. *Keidanren geppō* [Keidanren Monthly] 35(6): 35–38.

Kobayashi Shinichi. 1998. "Chishiki seisan system no hen'yō to science policy" [Transformation of Knowledge Production System and Science Policy]. *Bulletin of Institute for Higher Education* 16: 52–62.

Kobayashi Yoshinori. 1996. *Shin Gōmanism sengen special: datsu seigiron* [Neo Gōmanism Manifesto Special: Post-justice Theory]. Tokyo: Gentosha.

Kodama Kanichi. 1969. "Bunkyō Seisaku Iinkai chūkan hōkoku" [Interim Report of the Committee of Educational Policy]. *Journal of the JSEE* 17(1): 96.

Koga Issac. 1952. "Daigaku Denki Kyōkan Kyōgikai: tanjō no keii to sono dōsei" [The AAEEE: The Background of the Birth and Its Activities]. *Kaihō* [Bulletin of the JUAA] 11: 31–44.

———. 1973. "Watashi to Yoshida-kun tono kakawariai" [Relationship Between Mr. Yoshida and Me]. In Publishing Committee of Yoshida Goro's Memoirs 1973: 241–244.

Koga Issac, and Taki Yasuo. 1952a. "Daigaku no kōgyō kyōiku ni kansuru iken" [Opinions on Engineering Education at University]. *Journal of the IEEJ* 72(763): 1–3.

———. 1952b. "Corporation of Industrial Circle with Colleges in Engineering Education." [In Japanese.] *Journal of the IECEJ* 35(5): 1–2.

———. 1952c. "Engineering Education of the University." [In Japanese.] *Electrical Review* [Denki hyōron] 40(2): 2–6.

Kondo Kazuo. 1969. "Arubeki daigaku no sugata" [Ideal University]. In Moriguchi 1969: 134–143.

Kudo Hisha. 1995. "Gijutsushi to hon'yaku" [Gijutsushi and Translation]. *Gijutsushi: Consulting Engineer* 331 (September): 17–20.

Kuniya Minoru. 2015. *1980 nendai no kiso kagaku seisaku* [Basic Science Policy in the 1980s]. Tokyo: JISTEC.

Kuroda Kotaro. 1999. "Historical Review of Outlook on Industry-University Interactions in Japan." [In Japanese.] *Materia Japan* 38(11): 847–850.

Kusahara Katsuhide. 2012. "JABEE setsuritsu no zenshi" [Prehistory of JABEE]. In JABEE 2012: 15–19.

Layton, Edwin T., Jr. 1971. *The Revolt of the Engineers: Social Responsibility and the American Engineering Profession*. Cleveland: Press of Case Western Reserve University.

Lee, John A. N., and Jacques Berleur. 1994. "Progress Towards a World-Wide Code of Conduct." *ECA' 94: Proceedings of the Conference on Ethics in the Computer Age*, 100–104. New York: ACM.

Levitt, Theodore. 1983. "The Globalization of Markets." *Harvard Business Review* (May–June): 92–102.

Li Lihua. 2012. "Development of Industry-university Cooperation in Japan: Focusing on Its Major Characteristics by Each Period." [In Japanese.] *Bulletin of the Graduate School of Education, Hiroshima University. Part 3, Education and Human Science* 61: 233–242.

Maekawa Yoshihiro. 1987. "Jōhōka shakai no shinten to jōhō rinri" [Progress of Information Society and Information Ethics]. *Technology and Ethics in the Modern World: Nanzan Forum for Social Ethics* 3: 37–51.

Martin, Mike W., and Roland Schinzinger. 1983. *Ethics in Engineering*. New York: McGraw-Hill.

———. 1996. *Ethics in Engineering*. 3rd ed. New York: McGraw-Hill.

Matsumura Keiichi. 1950. *Beikoku sangyō jijō chōsa hōkokusho* [Survey Report on the Industry in the United States]. Postwar Economic Policy Materials of the Economic Stabilization Board, R16F3, microfilm of the Library of Economic Planning Agency collected by the Faculty of Economics, the University of Tokyo.

Mikami Yoshiki. 2006. "Gijutsusha rinri to gakukyōkai" [Engineering Ethics and Academic Societies]. *Journal of the Japan Fluid Power System Society* 37(2): 98–103.
Mikuriya Takashi. 2008. *Tōdai Sentanken monogatari* [A Story of RCAST]. Tokyo: RCAST.
Minagawa Masaru. 2009. "Results of the Questionnaire Survey on Engineering Ethics Education." [In Japanese.] Tokyo: JFES Council on Ethics for Engineers. December 2009. http://www.jfes.or.jp/_cet/topic/topic_no015_enquete.pdf.
Mitcham, Carl. 2001. "The Achievement of 'Technology and Ethics': A Perspective from the United States." In *Technology and Ethics: A European Quest for Responsible Engineering*, edited by Philippe Goujon and Bertrand H. Dubreuil, 565–581. Leuven: Peeters.
———. 2009. "A Historico-Ethical Perspective on Engineering Education: From Use and Convenience to Policy Engagement." *Engineering Studies* 1(1): 35–53.
———. 2019. *Steps Toward a Philosophy of Engineering: Historico-Philosophical and Critical Essays*. London: Rowman and Littlefield International.
MITI. 1957. *Sangyō gōrika hakusho* [The White Paper on Industrial Rationalization]. Tokyo: Nikkan Kogyo Shimbun.
Miyake Shigeru. 2000. "Kōgaku kyōiku kaikaku ni nozomu koto" [Request for Engineering Education Reform]. *STS Network Japan News Letter* 10(4): 2–3.
Miyamoto Takenosuke. 1932. "Doboku Gakkai kaizōron" [Reconstruction of the JCES]. *Doboku kōgaku* [Civil Engineering] 1(3): 62–63.
———. 1936. "Gijutsuka no shakaiteki danketsu" [Social Union of Engineers]. *Gijutsu Nippon* [Technology Japan] 165: 2–3.
———. 1937. "Kakushinteki kokusaku juritsu no yōken" [Requirements for Establishing Innovative National Policy] *Gijutsu Nippon* [Technology Japan] 176: 2–4.
———. 1941. "Gijutsusha undō no saishuppatsu" [Re-starting the Engineers' Movement]. *Gijutsu hyōron* [Technical Review] 18(1): 2–4.
Miyoshi Nobuhiro. 1999. *Tejima Seiichi to nihon kōgyō kyōiku hattatsushi* [Tejima Seiichi and the Development History of Engineering Education in Japan]. Tokyo: Kazama Shobo.
———. 2005. *Nihon kōgyō kyōiku hattatsushi no kenkyū* [Historical Research of the Development of Engineering Education in Japan]. Tokyo: Kazama Shobo.
Mizutani Masakazu. 1995. *Keiei rinrigaku no jissen to kadai* [Practice and Problems of Business Ethics]. Tokyo: Hakuto Shobo.
MOE. 1994. *Refresh kyōiku: shakaijin ni hirakareta daigaku guide (gakubu hen)* [Refresh Education: A University Guide to Adults (Undergraduate Edition)]. Tokyo: Gyosei.
Mori Wataru. 1989. *Sōchō shitsu no 1500 nichi* [1500 Days in the President's Office]. Tokyo: University of Tokyo Press.
Moriguchi Shigeichi, ed. 1969. *Atarashii kōgakubu no tameni* [For the New Faculty of Engineering]. Tokyo: University of Tokyo Press.
———. 1971. *Kōgakubu no kenkyū to kyōiku* [Research and Education of the Faculty of Engineering]. Tokyo: University of Tokyo Press.

Morihiro Eiichi. 2008. "PE Interview." [In Japanese.] *IPEJ Journal* 20(4): 16–19.
Morikawa Kakuzo. 1942. "Sōkan no ji" [Address for the First Issue]. *Nihon Nōritsu* [Japan Efficiency] 1(1): 1–3.
Morton, Rebecca J. 1983. *Engineering Law, Design Liability, and Professional Ethics: An Introduction for Engineers.* San Carlos, CA: Professional Publications.
Mukaibo Takashi. 1969. "Tōdai funsō no imisuru mono: gakubuchō no jinin ni atatte" [The Meanings of the Tōdai Riots: On the Resignation of the Dean]. In Moriguchi 1969: 118–123.
Murakami Yoichiro 1979. *Atarashii kagakuron: "jijitsu" wa riron o taoseruka* [A New Science Study: Can *Facts* Defeat Theory?]. Tokyo: Kodansha.
———. 1994a. *Kagakusha towa nanika* [What Is the Scientist?]. Tokyo: Shinchosha.
———. 1994b. "Kagaku gijutsu to rinri" [Science, Technology, and Ethics]. *FED Journal* 5(2): 1–4.
———. 1996. "Jōhō shori gijutsu to rinri" [Information Processing Technology and Ethics]. *IPSJ Magazine* 37(7): 671–678.
———. 2016. "Gakumon teki jiden" [Academic Autobiography]. In *MurakamiYoichiro no kagakuron: hihan to ōtō* [Science Studies of Murakami Yoichiro: Criticisms and Responses], edited by Kakihara Yasushi, et al., 15–69. Tokyo: Shinyosha.
Murakawa Jiro. 1981. "FIDIC eno kamei tassei no omoide" [Memories of the Achieving Accession to FIDIC]. In JCEA 1981: 90.
Nagao Makoto. 2010. *Jōhō o yomu chikara, gakumon suru kokoro* [The Ability to Read Information, the Mind to Study]. Kyoto: Minerva Shobo.
Nagaoka Hantaro. 1938. "Nihon Gakujutsu Shinkōkai: sōritsu gokanen to sono shōrai eno kibō" [The JSPSR: Five Years since Its Establishment and Hope for Its Future]. *Gakujutsu shinkō* [Promotion of Science] 8: 6–9.
Nakamura Shunichi. 1939. *Tsūshin gengyō jinji kanri* [Communications-Working-Personnel Management]. Tokyo: Kotsu Keizaisha.
Nakane Chie. 1970. *Japanese Society.* Berkley: University of California Press.
Namba Shogo. 1973. "Jitsuyōka to iu kotoba" [The Word Jitsuyōka]. In Publishing Committee of Yoshida Goro's Memoirs 1973: 398–400.
Nano Hiko. 1991. *Tōdai Sentanken: "sekai" o nerau "Nihon" no zunō* [RCAST: The Brain of Japan Aiming at the World]. Tokyo: NTT Publishing.
Nanzan University Institute for Social Ethics. 2006. *Shakai to rinri* [Society and Ethics] 20. Nagoya: Nanzan University.
Naoki Rintaro. 1918. *Gijutsu seikatsu yori* [From Technology Life]. Tokyo: Tokyodo.
Narita Hajime. 1997. *Pasokon hon'yaku no sekai* [The World of PC Translation]. Tokyo: Kodansha.
National Institute for Educational Research, ed. 1973. *Sangyō kyōiku 1* [Industrial Education 1]. Vol. 9 of *Nippon kindai kyōiku hyaku nenshi* [One-Hundred-Year History of Modern Education in Japan]. Tokyo: National Institute for Educational Research.
Natsume Kenichi. 2014. "Two Codes of Ethics Adopted by the Early Japan Consulting Engineers Association." [In Japanese.] *Journal of the Japan Society for the History of Industrial Technology* 19(1): 1–20.

———. 2015a. "Engineering Ethics and the International Consistency Problem of Gijutsushi in the 1990s: From Japan-U.S. Bilateralism to APEC Multilateralism." [In Japanese.] *Journal of the Japan Society for the History of Industrial Technology* 19(2): 17–41.

———. 2015b. "Statistics of Newspaper Articles about the Popularization of 'Engineering Ethics' in Japan in the 1990s." [In Japanese.] *Journal of the JSEE* 63(4): 53–58.

———. 2016a. "Historical Details of the Development of Engineering Ethics Education at the Kanazawa Institute of Technology in the 1990s." [In Japanese.] *Journal of the JSEE* 64(1): 39–44.

———. 2016b. "Engineering Professors and Industry-University Cooperation (1951–1969)." [In Japanese.] *Journal of the Japan Society for the History of Industrial Technology* 20(2): 11–19.

———. 2017a. "Promotion and Criticism of Industry-University Cooperation in Japan from the 1950s to 1960s." [In Japanese.] *Journal of Science and Technology Studies* 13: 32–47.

———. 2017b. "Yamakawa Kenjiro no kagaku shisō to shōbu shugi: butsurigaku, shakaigaku, fukoku-kyōhei." [Yamakawa Kenjiro's Scientific Thought and Pro-militarism: Physics, Sociology, and the Japanese Policy of Increasing Wealth and Military Power]. In *Meiji-Taishō ki no kagaku shisōshi* [Essays on the History of Scientific Thoughts in Modern Japan: Japanese Thoughts about Science Approximately Between the 1860s and 1930s], edited by Kanamori Osamu, 65–125. Tokyo: Keiso Shobo.

Nawa Kotaro. 1996. "Naze rinri kōryō ga hitsuyō ka" [Why Is the Code of Ethics Necessary?]. *IPSJ Magazine* 37(8): 793–794.

———. 1999. "Naze, ima jōhō rinri ka" [Why Information Ethics Now?]. *IPSJ Magazine* 40(11): 1119–1122.

Nawa Kotaro, and Otani Kazuko, eds. 2001. *IT user no hōritsu to rinri* [Laws and Ethics for IT Users]. Tokyo: Kyoritsu Shuppan.

Nawa Kotaro, and Yoneda Eiichi. 1996. "Rinri Kōryō Chōsa Iinkai" [The Code of Ethics Investigation Committee]. *IPSJ Magazine* 37(7): 702.

Nikkei Business Daily. 1994. "Gijutsusha, kyakkan hyōka ni michi" [Paving a Way for the Objective Qualifications for Engineers]. December 26, 1994.

Nikkei Shimbun. 1959a. "Kagaku Gijutsu Shinkō Zaidan: Keidanren de setsuritsu susumeru" [The Science and Technology Foundation: Keidanren Promotes the Establishment]. December 9, 1959.

———. 1959b. "*Sangaku-kyōdō center* setsuritsu" [The Establishment of the *Industry-Academia Cooperation Center*]. December 19, 1959.

———. 1986. "Tōkyō Daigaku, sentan gijutsu kenkyū, shin kikan: rainendo, bunka-kei tomo kōryū" [The University of Tokyo, Advanced Technology Research, New Institution: Next Year, Interact with Humanities]. July 7, 1986.

Nikkeiren [Japan Federation of Employers' Associations]. 1956. *Shinjidai no yōsei ni taiō suru gijutsu kyōiku ni kansuru iken* [Opinions on Technology Education Suitable for the Demands of the New Era]. Tokyo: Japan Federation of Employers' Associations.

Nishikiori Seiji. 1963. "Daidō Kōgyō Tanki Daigaku ni okeru sangaku-kyōdō" [Industry-Academia Cooperation at Daido Technical Junior College]. *Nihon no kagaku to gijutsu* [Japan Science and Technology] 43 (December): 14–16.

Nishino Fumio. 1997a. "Gijutsushi menkyo no kokusai sōgo nintei no ugoki to taiō (sono 1)" [The Movement and Response to the International Mutual Recognition of the Gijutsushi License (Part 1)]. *Sekisan gijutsu* [Cost Estimation Analysis] 231 (August): 28–33.

———. 1997b. "Gijutsushi menkyo no kokusai sōgo nintei no ugoki to taiō (sono 2)" [The Movement and Response to the International Mutual Recognition of the Gijutsushi License (Part 2)]. *Sekisan gijutsu* [Cost Estimation Analysis] 232 (September): 28–32.

———. 2004. "Kōeki wa rinriteki kōi no saiyūsen jikō" [Public Interest Is a Top Priority for Ethical Conduct]. *EAJ News* 97: 10–12.

Nishiyama Uzo, ed. 1978. *Ningen no songen to kagaku* [Human Dignity and Science]. Tokyo: Keiso Shobo.

Nitobe Inazo. 1905. *Bushido: The Soul of Japan*. revised ed. New York: G. P. Putnam's Sons.

———. 1919. "Heimindo" [The Citizens Way]. *Jitsugyō no Nihon* [Japan of Business] 22(10): 17–20.

Noguchi Isaaki. 1994. *Inoue Kowashi no kyōiku shisō* [The Educational Thoughts of Inoue Kowashi]. Tokyo: Kazama Shobo.

NRC. 1999. *Engineering Tasks for the New Century: Japanese and U.S. Perspectives*. Washington, DC: National Academies Press.

Ochi Mitsugu, Tutiya Syun, and Mizutani Masahiko, eds. 2000. *Jōhō rinrigaku: denshi network shakai no ethica* [Information Ethics: Ethica in the Electronic Network Society]. Tokyo: Nakanishiya Shuppan.

OECD. 1963. *The Measurement of Scientific and Technical Activities: Proposed Standard Practice for Surveys of Research and Experimental Development*. DAS/PD/62.47. 3rd revision. Paris: OECD. https://www.oecd.org/sti/inno/Frascati-1963.pdf.

———. 1996. *The Knowledge-Based Economy*. OECD/GD(96)102. Paris: OECD. http://www.oecd.org/officialdocuments/publicdisplaydocumentpdf/?cote=OCDE/GD(96)102&docLanguage=En.

Ogawa Kazuo. 1992. "Shōgai Iinkai ga hossoku shimashita" [The External Affairs Committee Was Established]. *Gijutsushi: Consulting Engineer* 292 (September): 25.

Ohashi Hideo. 1997. "Kōgaku Kyōiku Accreditation System Chōsa Kenkyū Iinkai shingi keika" [Progress of the Research Committee on the Accreditation System for Engineering Education]. *Journal of the JSEE* 45(2): 51–53.

Ohnaka Itsuo. 2000. "Nihon gijutsusha kyōiku nintei kijun no yōten to kyōiku no kaizen" [Essential Points of the Japanese Accreditation Criteria for Engineering Education and the Improvement of Education]. *Journal of the JSEE* 48(1): 21–25.

———. 2004. "Gijutsusha rinri ni taisuru JABEE no kenkai ni tsuite" [JABEE's View on Engineering Ethics], interview by Ishihara Koji, December 26, 2003. In *Kagaku gijutsusha rinri kyōiku system no chōsa kenkyū* [Research on the Science and Engineering Ethics Education System], edited by Nitta Takahiko, 307–310. Sapporo: Hokkaido University.

Okakura Koshiro. 1980. "UNESCO kankoku to kagakusha kenshō" [The UNESCO Recommendation and the Charter for Scientific Researchers]. In Watanabe and Igasaki 1980: 23–31.

Okamura Sogo. 1991. "Kōgaku kyōiku no kaikaku" [The Reform of Engineering Education]. *Journal of the JSEE* 39(3): 7–9.

———. 1999. "Sentan Gijutsu to Kokusai Kankyō Dai 149 Iinkai o tōshite no Beikoku Academy tono kōryū" [Interaction with the U.S. Academy through Committee 149 on Advanced Technology and the International Environment]. *Gakujutsu geppō* [Japanese Scientific Monthly] 52(3): 289–293.

Okazaki Tetsuji, Sugayama Shinji, Nishizawa Tamotsu, and Yonekura Seiichiro. 1996. *Sengo Nippon keizai to Keizai Doyukai* [Postwar Japanese Economy and Keizai Doyukai]. Tokyo: Iwanami Shoten.

Oki Michinori. 1978. "*Kagakusha Kenshō* ni tsuite" [On *the Charter for Scientific Researchers*]. In Nishiyama 1978: 1–19.

Okimoto, Daniel I. 1986. "The Japanese Challenge in High Technology." In *The Positive Sum Strategy: Harnessing Technology for Economic Growth*, edited by Ralph Landau and Nathan Rosenberg, 541–567. Washington, DC: National Academies Press.

Okochi Masatoshi. 1919. *Kōgyō kyōiku kanken* [Opinions on Engineering Education]. Tokyo: Okochi Masatoshi.

———. 1935. *Nōson no kōgyō* [Rural Industry]. Tokyo: Iwanami Shoten.

———. 1938. *Shihon shugi kōgyō to kagaku shugi kōgyō* [Capitalist Industry and Science-Based Industry]. Tokyo: kagaku Shugi Kōgyōsha.

Okoshi Takanori, ed. 1992. *Sentan kagaku gijutsu towa nanika* [What Is the Advanced Science and Technology?]. Tokyo: Asakura Publishing.

Onozawa Nagahide. 1981. "I. Gaiyō (kagaku seisaku no nichibei hikaku (I): dai 1 kai Nichibei Kagaku Seisaku Hikaku Kenkyū Seminar)" [I. Outline (U.S.-Japan Comparison of the Science Policies (I): The First U.S.-Japan Comparative Study Seminar on Science Policy)]. *Gakujutsu geppō* [Japanese Scientific Monthly] 34(3): 179–181.

Onuma Masanori, Fujii Yoichiro, and Kato Kunioki. 1975. *Sengo Nippon kagakusha undōshi* [The History of the Postwar Scientists' Movement], vol. 1. Tokyo: Aoki Shoten.

Osaka Higher Technical School. 1905a. "Osaka Kōtō Kōgyō Gakkō kinji" [News of Osaka Higher Technical School]. *The Educational Review* 725 (June): 32.

———. 1905b. "Osaka Kōtō Kōgyō Gakkō no kisoku kaisei" [Revision of the Rules of Osaka Higher Technical School]. *The Educational Review* 736 (September): 30.

———. 1911. *Osaka Kōtō Kōgyō Gakkō ichiran: Meiji 44–45* [Annual Report of Osaka Higher Technical School: FY 1911–1912]. Osaka: Osaka Higher Technical School.

Oshima Kiyoshi. 1935. "Tetsudō seishin ni taisuru ichi kōsatsu" [A Reflection on the Railway Spirit]. In *Tetsudō seishin kōza* [The Railway Spirit Course], edited by Tetsudō Dōyūkai, 189–207. Tokyo: Tetsudō Dōyūkai Honbu.

Oyama Matsujiro. 1950. "Beikoku no Professional Engineer ni tsuite" [About the American Professional Engineer]. *Journal of the IEEJ* 70(740): 205.

———. 1952. "Monbu Daijin sonota ni taisuru 'kōgyō kyōiku sinkō chinjōsho" [A Petition for Promotion of Engineering Education to the MOE Minister and Others]. *Journal of the JSEE* 1(1): 91–93.

Oyodo Shoichi. 1989. *Miyamoto Takenosuke to kagaku gijutsu gyōsei* [Miyamoto Takenosuke and the Administration of Science and Technology]. Tokyo: Tokai University Press.

———. 2009. *Kindai Nippon no kōgyō rikkokuka to kokumin keisei: gijutsusha undō ni okeru kōgyō kyōiku mondai no tenkai* [The Industry-Oriented National Development of Modern Japan and the Formation of the Nation: Development of the Problems of Engineering Education on the Engineers' Movement]. Tokyo: Suzusawa Shoten.

Ozeki Masanori, et al. 1995. "Jōhōka shakai no kihan" [Information Ethics]. *IPSJ Magazine* 36(11): 1074–1079.

Pacific Consultants. 2002. *Mirai o umu rekisi* [History to Create the Future]. Tokyo: Pacific Consultants Group.

Phillips, Winfred M., George D. Peterson, and Kathryn B. Aberle. 2000. "Quality Assurance for Engineering Education in a Changing World." *International Journal of Engineering Education* 16(2): 97–103.

Polkinghorn, Frank A. 1949a. "Some Observations Concerning Engineers and Engineering in Japan." *Journal of the IECEJ* 32(4): 1–4. Original English text in Polkinghorn 1969.

———. 1949b. "Remarks at the Formal Opening of the Electrical Communication Laboratory." *The Monthly Journal of the ECL* 2(6): 219–222.

———. 1949c. "The Organization of Research & Development." *Journal of the IEEJ* 69(7): 1–5.

———. 1950a. "A Talk to Future Communications Engineers." *Electrical Review* [Denki hyōron] 38(8): 2–5.

———. 1950b. "The Improvement of Telecommunications Engineering Education in Japan." Trans. Koga Issac. *Journal of the IECEJ* 33(10): 1–4. Original English text in Polkinghorn 1969.

———. 1952. "Forward to the CCS Book on Industrial Management." [In Japanese.] In Vol. 1 of *The CCS Management Course*, edited and translated by the FJECIA, 7–11. Tokyo: Diamond. Original English text in Polkinghorn 1969.

———. 1969. *Diary & Memoirs of Two Years with the Occupation in Japan*. Box 3, Kenneth Hopper Papers on Management, Drucker Institute. http://ccdl.libraries.claremont.edu/cdm/search/collection/khp/searchterm/khp00199.pdf.

———. n.d.a. "Association of the Advancement of Electrical Engineering Education." Box 18, Folder 30, Kenneth Hopper Papers on Management, Drucker Institute. http://ccdl.libraries.claremont.edu/cdm/search/collection/khp/searchterm/khp00815_0001.tif.

———. n.d.b. "Schedule of Major Activities with Telecommunications Professors." Box 18, Folder 32, Kenneth Hopper Papers on Management, Drucker Institute. http://ccdl.libraries.claremont.edu/cdm/search/collection/khp/searchterm/khp00842.

Postwar Communications Industry Editorial Committee, ed. 1959. *Sengo no tsūshin kōgyō* [Postwar Communications Industry]. Tokyo: FJECIA.
Preparatory Investigation Committee for the University Reform. 1969. *Tokyo Daigaku kaikaku junbi chōsakai hōkokusho* [Report of the Preparatory Investigation Committee for the Reform of the University of Tokyo]. Tokyo: University of Tokyo Press.
Production Control Committee, Ministry of Commerce and Industries. 1938. *Kōgyō kyōiku o chūshin toshite mita wagakuni kyōiku seido no kaizen* [Improvement of Our Country's Education System with a Focus on Engineering Education]. Tokyo: Nihon Kōgyō Kyōkai.
Provisional Council on Education Reform. 1988. *Kyōiku kaikaku ni kansuru tōshin: Rinji Kyōiku Shingikai dai 1–4 ji (saishū) tōshin* [Reports on Educational Reform: Provisional Council on Education Reform First to Fourth (Final) Reports]. Tokyo: National Printing Bureau.
Public Relations Office, University of Tokyo. 1987. *Gakunai kōhō* [Campus Newsletter]. 754 (June 1).
Publishing Committee of Yoshida Goro's Memoirs, ed. 1973. *Yoshida Goro no omokage* [Yoshida Goro's Memoirs]. Tokyo: Yoshida Goro Tsuitōroku Kankōkai [Publishing Committee of Yoshida Goro's Memoirs].
Putnam, Robert D. 2000. *Bowling Alone: The Collapse and Revival of American Community*. New York: Simon & Schuster.
RCAST. 2007. *Tokyō Daigaku Sentan Kagaku Gijutsu Kenkyū Center nijū nenshi: aru ichi bukyoku no jishōroku* [Twenty-Year History of RCAST: A Self-Reflection of a Certain Department]. Tokyo: RCAST.
Record Editorial Committee for the Waseda Struggle. 1966. *Waseda o yurugashita 150 nichi: Sōdai tōso no kiroku* [150 Days of Jolting Waseda: Record of the Waseda Struggle]. Tokyo: Gendai Shobo.
Reed, Robert F. 1983. *The US-Japan Alliance: Sharing the Burden of Defense*. Washington, DC: National Defense University Press.
Research Committee on Refresh Education in Engineering. 1993. *Kōgakukei bun'ya ni okeru refresh kyōiku no arikata ni kansuru chōsa kenkyū hōkokusho* [Report of the Research Committee on Refresh Education in Engineering]. Tokyo: JSEE.
RIITI. 1980. *80 nendai no tsūshō seisaku vision* [Trade Policy Vision in the 1980s]. Tokyo: RIITI.
RIKEN. 1917. "Rikagaku Kenkyūsho setsuritsu no shushi narabini setsuritsu keikaku no taiyō" [Purpose of the Establishment of RIKEN and the Outline of the Establishment Plan]. June 1917.
Saeki Kazuyoshi. 2007. "Taki Shigeatsu-san o shinonde" [In Memory of Mr. Taki Shigeatsu]. In Umeda 2007: 112–114.
Sakurai Yoshiko. 1994. *AIDS hanzai: ketsuyūbyō kanja no higeki* [AIDS Crime: Tragedy of the Hemophiliac Patients]. Tokyo: Chuokoron.
Sano Toshikata, et al. 1936. "Zadankai: gijutsusha no shakaiteki chii ninmu ni tsuite" [Round-Table Discussion: The Social Status and Mission of Engineers]. *Gijutsu Nippon* [Technology Japan] 169: 24–46, 48.
Sasaki Shigeo. 1951. "Kōgyō Kyōiku Kenkyū Shūkai no gaikyō" [Overview of the IFEE]. *Kaihō* [Bulletin of the JUAA] 10: 27–36.

———. 1962. "Yoakemae no omoide" [Memoirs before Dawn]. *Journal of the JSEE* 10(1): 29–35.
Sato Fumitaka, and Takeuchi Kei. 1996. "Taidan: kagakusha wa dokoni iruka" [Interview: Where Are the Scientists]. *Gendai shisō* [Revue de la pensée d'aujourd'hui] 24(6): 52–68.
Sato Kiyoshi. 2002. "Yakuhenja jo" [Preface by the Translators and Editors]. In Harris, Jr., et al., 2000: III–V.
———. 2007. "Taki-san no ningen ryoku" [Mr. Taki's Ability]. In Umeda 2007: 50–51.
Sato Mitsuharu. 1979. "Rinri ni kibishii America Gasshū Koku no doboku kai" [Civil Engineering Society in the United States, Which Is Strict in Ethics]. *Journal of the JSCE* 64(9): 2–9.
Sawada Yoshiro. 2011. "Sangaku-renkei, chiteki zaisan seisaku no tenkai to kokuritsu daigaku no konran" [The Development of Industry-Academia Collaboration and Intellectual Property Policy and the Confusion in National Universities]. In Vol. 3 of *Shin tsūshi: Nihon no kagaku gijutsu* [New Comprehensive History: Science and History in Japan], 120–126. Tokyo: Hara Shobo.
Sawai Minoru. 2012. *Kindai Nippon no kenkyū kaihatsu taisei* [Research and Development System of Modern Japan]. Nagoya: Nagoya University Press.
Schneider, Herman. 1906. "Technical Education for Cincinnati." *University of Cincinnati Record* ser. 1, 2(9): 3–9.
SCJ. 1972. *1970 nendai ikō no kagaku gijutsu ni tsuite* [Science and Technology since the 1970s]. Tokyo: National Printing Bureau.
———. 1980. *Kagakusha Kenshō ni tsuite* [On the Charter for Scientific Researchers]. Tokyo: SCJ. http://www.scj.go.jp/ja/info/kohyo/09/11-18-s.pdf.
———. 1996. "Gakujutsu jōhō hasshin kichi: gakujutsu dantai no kyōka shien ni mukete" [Academic Information Center: To Strengthen and Support Academic Organizations]. *Gakujutsu no dōkō* [Trend in the Sciences] 1(8): 58–65.
———. 1997. *Kōgakukei kōtō kyōiku kikan deno gijutsusha no rinri kyōiku ni kansuru teian* [Proposal for Ethics Education for Engineers at Higher Education Institutions in Engineering]. Tokyo: SCJ Section 5.
———. 2010. *Teigen: Nihon no kiso kagaku no hatten to sono chōki tenbō* [Proposal: Development of Basic Science in Japan and Its Long-Term Perspective]. Tokyo: SCJ. http://www.scj.go.jp/ja/info/kohyo/pdf/kohyo-21-tsoukai-5.pdf.
SCM. 1973. "Gakujutsu shinkō ni kansuru tōmen no kihonteki na shisaku ni tsuite" [Immediate Basic Policies for Academic Promotion: the Third SCM Report]. SCM 10 (October 31, 1973) In *Gakujutsu geppō* [Japanese Scientific Monthly] 26, special issue 3, 1974: 1–114.
———. 1984 "Gakujutsu kenkyū taisei kaizen no tameno kihonteki shisaku ni tsuite (tōshin)" [Basic Measures for Improvement of Academic Research System (Report)]. *Gakujutsu geppō* [Japanese Scientific Monthly] 36(11): 33–58.
Shibata Shingo. 1966. "Dokusen shihon no fukkatsu to daigaku" [Revival of Monopoly Capital and University]. *Journal of Japanese Scientist* 1(1): 34–39.
Shimizu Kinji. 1952. "Beikoku no kōgyō kyōiku (3)" [American Engineering Education (3)]. *Kaihō* [Bulletin of the JUAA] 14: 18–32.

———. 1956. "Dai 4 kai nenji taikai kaichō aisatsu" [President's Greeting at the Fourth Annual Conference]. *Journal of the JSEE* 4(1–2): 3–8.

———. 1960. "Shimon dai 1 gō ni kansuru shingi ni sanka shite" [Participated in Deliberations on the Advisory Report No. 1]. *Gakujutsu geppō* [Japanese Scientific Monthly] 13(7–8): 102–103.

———. 1962. "Nihon Kōgyō Kyōiku Kyōkai oyobi kaku chiku Kōgyō Kyōiku Kyōkai no oitachito sononochi: jūnen o kaerimiru" [The History of the JSEE and the Regional Societies for Engineering Education: Looking Back on the Ten Years]. *Journal of the JSEE* 10(1): 8–19.

Shiota Eizo [Hirayama Fukujiro]. 1951. "Nihon Gijutsushi-kai no hossoku" [The Launch of the JCEA]. *Journal of the JSCE* 6(12): 47–49.

Shiraishi Shunta. 1994. *Shiraishi Tashirō ryakuden* [Biographical Sketch of Shiraishi Tashiro]. Tokyo: Tashi Fudosan.

Spencer, Herbert. 1864–1867. *The Principles of Biology*. 2 vols. New York: D. Appleton.

STA. 1963. *Kigyō no kenkyū katsudō ni kansuru chōsa* [Survey on Corporate Research Activities]. Tokyo: STA.

STS Network Japan. 2002. "Dai 24 kai symposium: kōgaku kyōiku to STS no kanōsei" [The Twenty-Fourth Symposium: Possibilities of Engineering Education and STS]. *STS Network Japan Yearbook* 9: 32–47.

Subcommittee of Japan Scientists' Association in Faculty of Engineering at Kyoto University. 1970. "Kōgakubu ni okeru sangaku-kyōdō" [Industry-Academia Cooperation in the Faculty of Engineering]. *Journal of Japanese Scientist* 5(2): 16–23.

Suga Hidemi. 1998a. "Muishiki to shite no *sangaku-kyōdō hantai* (1)" [*Objection to the Industry-Academia Cooperation* at Unconscious (1)]. *Hatsugensha* [Monthly Speak-out Magazine] 49: 100–105.

———. 1998b. "Muishiki to shite no *sangaku-kyōdō hantai* (3)" [*Objection to the Industry-Academia Cooperation* at Unconscious (3)]. *Hatsugensha* [Monthly Speak-out Magazine] 51: 104–109.

Sugihara Keita. 2007. "A Study of Engineering Ethics Integrated with Science and Technology Studies." [In Japanese.] PhD Dissertation, Nagoya University.

Sugimoto Taiji. 1995. "Gijutsushi no jōkyō ni kansuru iken" [An Opinion on the Situation Around Gijutsushi]. *Gijutsushi: Consulting Engineer* 327 (June): 13–16.

Sugimoto Taiji, and Taki Shigeatsu. 2001. *Daigaku kōgi: gijutsusha no rinri nyūmon* [University Lecture: Introduction to Ethics of Engineers]. Tokyo: Maruzen.

———. 2002. *Daigaku kōgi: gijutsusha no rinri nyūmon* [University Lecture: Introduction to Ethics of Engineers]. 2nd ed. Tokyo: Maruzen.

Suguri Yukiyasu. 1999. "Ideology of IEEE-SIT Tokyo and the Present Situation." [In Japanese.] *IEICE Technical Report* 99(348): 57–62.

Tachi Akira. 1983. "Kigyō to daigaku: senzen no sobyō" [Companies and Universities: A Prewar Sketch]. *IDE: Gendai no kōtō kyōiku* [IDE: Modern Higher Education] 244: 5–11.

Takada Ichiro. 1958. "Gijutsushi-kai no tanjō tōji no kiroku" [The Birth of the JCEA: A Record of Those Days]. *JCEA* 24: 11–13.

Takahashi Osamu. 2007. "Taki Shigeatsu-shi o shinobu" [In Memory of Mr. Taki Shigeatsu]. In Umeda 2007: 178–179.
Taketani Mitsuo. 1961. "Sangaku-kyōdō ron no shisei o tadasu" [Correct the Attitude of the Discussion of Industry-Academia Cooperation]. *Economist* 28 (July): 42–47.
Taki Shigeatsu. 1991. "Nihonjin demo shutoku dekiru Beikoku no P. E." [American PE that Even Japanese Can Obtain]. *Gijutsushi: Consulting Engineer* 280 (November): 28.
———. 1993. "Gijutsushi shikaku seido no kokusai seigōsei" [International Consistency of the Gijutsushi System]. *Gijutsushi: Consulting Engineer* 302 (June): 34–37.
———. 1994. "Engineer sikaku no kokusaika o sasaeru yōken" [Requirements to Support the Internationalization of Engineering Qualifications]. *Gijutsushi: Consulting Engineer* 315 (June): 62–64.
———. 1995. "Jizokuteki hatten ni okeru gijutsusha no yakuwari" [The Role of Engineers in Sustainable Development]. *Gijutsushi: Consulting Engineer* 329 (June): 49–51.
Tanaka Hiroshi. 1950a. *Beikoku ni okeru kikai kōgyō ni kansuru chōsa hōkokusho: tobei hōkokusho* [Survey Report on the Mechanical Industry in America: A Report on the United States]. Postwar Economic Policy Materials of the Economic Stabilization Board, R16F3, microfilm of the Library of Economic Planning Agency collected by the Faculty of Economics, The University of Tokyo.
———. 1950b. "Beikoku no consulting engineer ni tsuite" [About Consulting Engineer in America]. *Nissankyō geppō* [Monthly Report of Japan Industry Council] 5(12): 18–19.
———. 1951a. *Beikoku ni okeru consulting engineer ni tsuite* [About Consulting Engineer in America]. Tokyo: Nihon Sangyo Kyogikai.
———. 1951b. "America ni okeru consulting engineer no jittai: meikaku na ryōkin seido, genkaku na shikaku, seijō na rinrikan" [Actual Situation of Consulting Engineers in America: Clear Price System, Strict Qualification, and Proper Sense of Ethics]. *Management* 10(5): 13–17.
———. 1951c. "Consulting Engineer no hattensei ni tsuite" [On the Possibilities of Consulting Engineer]. *Industrial Technology* 1(6): 2–6.
Taylor, Frederick W. 1911. *Principles of Scientific Management*. New York: Harper & Brothers.
———. 1947. *Taylor's Testimony before the Special House Committee*. New York: Harper & Row.
Technical Education Division of the MOE. 1952. *Report of the Engineering Education Mission to Japan*. Tokyo: Technical Education Division of the MOE.
Tejima Seiichi. 1909. "Shōnindō to Bushidō" [The Merchant Way and the Samurai Way]. *Nihon oyobi Nihonjin* [Japan and the Japanese] 517: 24–28.
TEPCO. 2002–2015. "Holding of the Corporate Ethics Committee." [In Japanese.] 44, 55, 56, 58, 61, 62, 63, 64, 65. Tokyo: TEPCO. http://www.tepco.co.jp/csr/kai fuku/results-j.html.

———. 2014. "Holding of the Sixty-Fifth Corporate Ethics Committee." [In Japanese.] Tokyo: TEPCO. http://www.tepco.co.jp/csr/kaifuku/trend/rinri65-j.html.

Tezuka Akira. 1981. "B. Nihon ni okeru kagaku gijutsu seisaku keisei no kihon kōzō to jittai (Kagaku seisaku no nichibei hikaku (I): II. Kagaku seisaku no wakugumi)" [B. Basic Structure and Actual Situation of the Science and Technology Policy-Making in Japan (U.S.-Japan Comparison of Science Policy (I): II. The Framework of the Science Policy)]. *Gakujutsu geppō* [Japanese Scientific Monthly] 34(3): 186–188.

Tokai Society for Engineering Education. 1955. Workshop Report. [In Japanese.] *Journal of the JSEE* 3(1): 101–159.

Tokyo Chamber of Commerce and Industry. 1973. *Shinjidai ni sokuō suru sangaku-kyōdō no arikata ni kansuru iken chōsa* [Opinion Survey on the Way of Industry-Academia Cooperation to Respond Promptly to the New Era]. Tokyo: Tokyo Chamber of Commerce and Industry.

Tokyo Institute of Technology, ed. 1940. *Tokyo Kōgyō Daigaku rokujū nenshi* [Sixty-Year History of Tokyo Institute of Technology]. Tokyo: Tokyo Institute of Technology.

Tokyo Technical School. 1899. *Tokyo Kōgyō Gakkō ichiran: Meiji 32–33* [Annual Report of Tokyo Technical School: FY 1899–1900]. Tokyo: Tokyo Technical School.

Toyo University. 1994. *Tōyō Daigaku hyaku nenshi: tsūshi hen II* [One-Hundred-Year History of Toyo University: Comprehensive History II]. Tokyo: Toyo University.

Tsujii Shigeo. 1995. "On the Study of Info-Communication Ethics: Revolution of Culture and Civilization Structure by Personalization and Globalization of Information Network." [In Japanese.] *IEICE Technical Report* 95(64): 1–6.

Tsuzaki Masanosuke. 1951. *America no shokugyō shidō to shokugyō kyōiku* [Vocational Guidance and Vocational Education in America]. Tokyo: Bunkyoshoin.

———. 1972. "Nihon Kōgyō Kyōiku Kyōkai no kongo" [The Future of the JSEE]. *Journal of the JSEE* 19(2): 2–4.

Uchida Kozo. 1996. *Keidanren to Nippon keizai no 50 nen* [Fifty Years of Keidanren and Japanese Economy]. Tokyo: Nikkei.

Uchida Nobuya. 1935. "Tetsudō seishin" [The Railway Spirit]. In *Tetsudō seishin kōza* [The Railway Spirit Course], edited by Tetsudō Dōyūkai, 1–8. Tokyo: Tetsudō Dōyūkai Honbu.

Uemura Yukio. 1989. *Kagaku gijutsu seisakuron* [A Study of Science and Technology Policy]. Tokyo: Rodo Junposha.

Ui Jun. 1985. "Saraba Tōdai: (1) goyō-gakusha tono tatakai" [Goodbye, Tokyo University: (1) Battle with the Obsequious Scholars]. *Asahi Journal* 27(50): 22–26.

Umeda Masao. 1998. "Yakuhenja jo" [Preface by the Translators and Editors]. In Harris, Jr., et al. 1998: III–IV.

———, ed. 2007. *Taki Shigeatsu-san tsuitōroku* [Memoirs of Mr. Taki Shigeatsu]. Tokyo: Umeda Masao.

Unayama Hideo, and Ogura Ichiro, eds. 1996. *Kigyō rinri to kaikei fusei* [Corporate Ethics and Account Fraud]. Tokyo: TEIP.

University of Chicago Press. 2017. *The Chicago Manual of Style*. 17th ed. Chicago: University of Chicago Press.
University of Tokyo Faculty and Staff Union. 1987. "Kifu kōza: daigaku ni nani o motarasuka" [Endowed Chairs: What Will Be Brought to the University?]. *Forum: Tōdai kaikaku* [Forum: The University of Tokyo Reform]. 2 (October 14).
Utsumi Kiyoharu, Hiki Hajime, et al. 1966. "Gijutsushi-kai no nenrin: omoide to genjō o kataru (zadankai)" [Annual Ring of the JCEA: Talk about Memories and the Current Situation (Round-Table Talk)]. *Consultant* 17: 30–38.
Wada Masao. 1956. "Nenji taikai no omoide" [Memories of the Annual Conference]. *Journal of the JSEE* 4(1–2): 218–221.
———. 1969. "Kōgakukei gakusotsusha ni taisuru sangyōkai no yōbō ni kotaeru michi: sangaku-kyōdō o suishin seyo" [The Way to Meet the Demands of the Industry for Engineering Graduates: Promote Industry-University Cooperative Education]. *Journal of the JSEE* 16(2): 34–39.
Watanabe Hiromoto. 1893. "Gijutsusha sekinin ni tsuite" [On the Responsibility of Engineers]. *Journal of the Engineering Society* 133: 4–15.
Watanabe Naotsune, and Igasaki Akio, eds. 1980. *Kagakusha Kenshō* [The Charter for Scientific Researchers]. Tokyo: Keiso Shobo.
Watanabe Shigeru. 1969. "Taishūka suru daigaku" [University in Mass Popularization]. In Moriguchi 1969: 143–155.
Weil, Vivian. 1984. "The Rise of Engineering Ethics." *Technology in Society* 6: 341–345.
WFEO Committee on Education and Training. 1994. *Ideas: For Better Education & Training for Engineers* 2. December 1994.
Widegren, Ragnar. 1988. *Consulting Engineers 1913–1988: FIDIC Over 75 Years*. Stockholm: International Federation of Consulting Engineers.
Wittrock, Björn, and Michael Gibbons, eds. 1985. *Science as a Commodity: Threats to the Open Community of Scholars*. Essex: Longman.
Wolferen, Karel van. 1994. *Ningen o kōfuku ni shinai Nippon toiu system* [The False Realities of a Politicized Society]. Translated by Shinohara Masaru. Tokyo: Mainichi Newspapers.
Yagi Hidetsugu. 1926. "Ningen toshite no kyōiku wa mottomo jūyō nari" [Human Education Is the Most Important]. In *Kōgyō kyōiku no kenkyū* [Research on Engineering Education], edited by Koseikai, 52–53. Tokyo: Koseikai Shuppanbu.
Yamagami Susumu, ed. 1994. *Asia-Taiheiyō chiiki no jidai: APEC setsuritsu no keii to tenbō* [The Age of the Asia-Pacific Region: History and Prospects of the Establishment of APEC]. Tokyo: Dai-ichi Hoki.
Yamaguchi Yoshikazu, and Shibata Kiyoshi. 2017. "Change of the Themes Related to Engineering Ethics Based on the Analysis of Newspaper Articles." [In Japanese.] *Journal of the JSEE* 65(6): 49–55.
Yamakawa Kenjiro. (1905) 1937a. "Teikoku Daigakurei kōfu kinen shukuga shiki enzetsu" [Speech at the Celebration Ceremony for the Promulgation of the Imperial University Order in March 1905]. In Ko-Yamakawa Danshaku Kinenkai 1937: 187–193.

———. 1911. "Kunji (Meiji yonjūninen shigatsu tsuitachi kari-kaikōshiki ni oite)" [Instruction (at the Temporary Opening Ceremony on April 1, 1909)]. In *Shiritsu Meiji Senmon Gakkō ichiran: Meiji 43–44* [Annual Report of MCT: FY 1910–1911], 50–63. Fukuoka: MCT.

———. (1912) 1937b. "Daiikkai sotsugyōshiki ni okeru kokuji (Meiji yonjūgonen shichigatsu)" [Announcement at the First Graduation Ceremony in July 1912]. In Ko-Yamakawa Danshaku Kinenkai 1937: 242–245.

———. (1913) 1937c. "Dainikai sotsugyōshiki ni okeru kokuji: Rokoku ni okeru Yudayajin gyakutai (Taisho ninen)" [Announcement at the Second Graduation Ceremony: The Jewish Persecution in Russia (1913)]. In Ko-Yamakawa Danshaku Kinenkai 1937: 245–251.

———. (1929) 1937d. "Rokujūnen mae gaiyū no omoide" [Memories of the Overseas Travel Sixty Years Ago]. In Ko-Yamakawa Danshaku Kinenkai 1937: 42–73.

Yoneda Eiichi. 1997. "*Internet to rinrikan* ni kansuru zakkan" [Impressions on *Internet and Ethics*]. *Bit* 29 (October): 10–14.

Yoshida Goro. 1948a. "Hakkan no ji" [Publication Address]. *The Monthly Journal of the ECL* 1(1): 1.

———. 1948b. "Yoshida shochō aisatsu" [Greetings from Director Yoshida]. *The Monthly Journal of the ECL* 1(1): 2–4.

———. 1948c. "Teishinshō Denki Tsūshin Kenkyūjo seitan no hitsuzensei to sono ninmu" [The Necessity of the Birth of the ECL and Its Mission]. *The Monthly Journal of the ECL* 1(1): 12–31.

———. 1965. "Denki Tsūshin Kenkyūjo sōritsu no sōsetsu tōji ni omoi o hasete" [Thinking back when the ECL Was Established]. In *Denki Tsūshin Kenkyūjo jūgo nenshi* [Fifteen-Year History of the ECL], edited by Fifteen-Year History of the ECL Editorial Committee. Tokyo: ECL.

———. 1973. "Denki Tsūshin Kenkyūjo sōritsu no seishin ni tsuite" [About the Founding Spirit of the ECL]. In Publishing Committee of Yoshida Goro's Memoirs 1973: 88–124.

Yoshikawa Hiroyuki. 1990. *Gainen no sekkei kara shakai system e* [From Conceptual Design to Social System]. Tokyo: Mita Shuppankai.

———. 1997. "Kagakusha no rinri: dai 17 ki Nihon Gakujutsu Kaigi Kaichō ni shūnin shite" [Ethics of Scientists: On Being Appointed as the Seventeenth SCJ President]. *Gakujutsu no dōkō* [Trend in the Sciences] 2(9): 5–7.

———. 1998. "Nentō shokan: Nihon byō to shinnen no yume" [A New Year's Opinion: Japanese Disease and a New Year's Dream]. *Gakujutsu no dōkō* [Trend in the Sciences] 3(1): 5–7.

Yoshimura Masamitsu. 1952. "Gijutsushi no hōhōron" [Methodology of Gijutsushi]. *Chemistry and Chemical Industry* 5(1): 11–17.

Yoshino Sakuzo. 1916. "Kensei no hongi o toite sono yūshū no bi o nasuno michi o ronzu" [Discussing the True Meaning of Constitutional Government to Complete the Final Beauty]. *Chuokoron* 31(1): 17–114.

Index

AAEEE. *See* Association for the Advancement of Electrical Engineering Education
academic freedom, 32n12, 51, 57n11, 85, 108
Accreditation Board for Engineering and Technology (ABET), 5, 111–13, 135, 138, 147–48, 150, 161, 163, 171n8
American Institute of Consulting Engineers (AICE), 4, 21, 66–67, 70–71, 182
American Institute of Electrical Engineers (AIEE), 4, 33n24, 48–49. *See also* Institute of Electrical and Electronics Engineers
American Society for Engineering Education (ASEE), 30, 35–36, 51–52, 55, 59n26, 184–85
American Society of Civil Engineers (ASCE), 4, 9, 21, 25–28, 33nn24–25, 99, 171n4, 175
American Society of Mechanical Engineers (ASME), 4, 33n24, 100, 161
Anan Seiichi, 116
Aoki Seizo, 90
applied ethics, 5, 7, 114–19, 178
Applied Ethics Center for Engineering and Science (ACES). *See* Kanazawa Institute of Technology

Asahara Genshichi, 65–66
Asia-Pacific Economic Cooperation (APEC), 1, 131–39, 141–43, 144n7, 144n11, 144n13, 145n16, 145n18, 146n29, 147–48, 152–53, 166–69, 170n2, 171n8, 171n10, 177–78, 181
Association for the Advancement of Electrical Engineering Education (AAEEE), 3–4, 36, 44, *50*, 185
Association of Consulting Engineers (ACE), 70
Association of Japanese Consulting Engineers (AJCE), 72, 74, 78n16, 128, 152
Aum Shinrikyo, 2, 7n2, 122, 157, 159, 167, 177
autonomy, 6, 13, 16, 18–19, 41, 61, 73, 75–76, 80, 85–88, 91, 96n9, 99–100, 103, 108, 110, 117–18, 148–49, 151, 159, 161–62, 166–70, 178–79

Basic Act on Science and Technology, 85, 98, 106, 159, 171n9
Basic Act on Scientific Research, 85, 98
basic research, 22, 37, 83, 91, 101–7, 109–11, 123nn7–8, 124n11, 124n13, 176; free ride in, 106–7, 109
Bayh-Dole Act, 104

Bell Telephone Laboratories, 38–41, 107
bioethics, 5, 115–16, 119, 125n24
Boy Scout, 12
brain death, 115, 157–58
Bushido, 3, 12, 14–16, 20, 32n7, 32n9, 47, 57n9, 100, 175, 181, 185. *See also* Confucianism; loyalty

Civil Communications Section (CCS), 36–44, *45*, 49, 58n16, 61, 176; Industry Division of, 36–37, 41; Research and Development Division of (CCS/R&D), 36–37, 39, 41, 43, 49; Telephone and the Telegraph Division of (CCS/T&T), 43–44. *See also* General Headquarters, Supreme Commander for the Allied Powers
Civil Information and Educational Section (CI&E), 3, 35, 44, *45*, 57n11, 61, 184. *See also* General Headquarters, Supreme Commander for the Allied Powers
Civil Service Appointment Ordinance, 21–22, 33n16
Civil War, 14–15
cloning technology, 157
collectivism and conformism, 2, 145n20, 154–55, 178
college riots, 85, 87, 91–92, 102, 107–9, 124n18, 172n19, 186. *See also* student activism
commissioned research, 8n10, 54–55, 82–84, 89–90, 94, 96n6
communism, 18–19, 24, 28, 37, 57n11, 97, 123n4
confidentiality, 54–55, 66–67, 70, 76
Confucianism, 10, 100, 181, 185. *See also* Bushido; loyalty
Consulting Engineer (CE), 1, 4, 61–76, 77nn6–7, 77n9, 77–78nn12–14, 78n17, 127–31, 133, 136–38, 144n7, 167, 176–77, 182
continuing professional development (CPD), 134, 145n16

Coordinating Committee for Multilateral Export Control (CoCom), 78n18, 118, 141
corporate ethics, 5, 118–19, 155, 169, 172n11, 177
Council for Science and Technology (CST), 81, 83, 85, 103, 123n8, 189

Dees, Bowen C., 35, 51
democracy, 3, 6–7, 19–20, 23–25, 28–29, 31, 35–37, 41–43, 46–49, 56, 57n11, 58n16, 59n27, 61, 75–76, 79, 85–87, 95, 96n9, 97–99, 109–10, 151–52, 158–60, 170, 175–79, 188–89; Taisho Democracy, 19–25, 28–29, 175
democratization: of education, 3, 35, 47–49, 57n11, 87, 95, 99, 109–10, 151–52, 175–76; of industry, 36, 75; of management, 40–42; of service, 37, 40–41; of social system, 35, 41–42, 46–47, 56, 75, 159–60, 177, 188
dignity, 3, 26, 42, 66–67, 70, 74, 76
Drucker, Peter, 113
Dyer, Henry, 9–10

Economic Stabilization Board, 63–65, 68, 73
Eells, Walter C., 44, *45*, 57
Eight-University Council of Deans of the Engineering Faculties, 125n20, 170n1
Electrical Communication Laboratory (ECL), 37, 39–40, 43, *45*, 47, 56n1
Electrotechnical Laboratory (ETL), 37, 39, 46–47, 51, 146n32
emperor, 10–11, 13, 19–21, 24, 28, 31n2, 185, 189. *See also* Kokutai
Engineering Academy of Japan (EAJ), 112, 124n12, 165, 183, 186
Engineering Advancement Association of Japan (ENAA), 143n2
Engineering Council (EngC), 111, 171n8
Engineering Criteria 2000 (EC2000), 5, 150, 161

Engineering Education Mission to Japan, 30, 35, 51–52, 57–58n12, 58n22, 112, 184–85. *See also* Japan Federation of Engineering Societies
Engineering Society, 3, 22, 183.
Engineers' Council for Professional Development (ECPD), 4–5, 48, 54, 66–67, 70, 111. *See also* Accreditation Board for Engineering and Technology
environment: conservation of, 137, 161, 168; ethics of, 5, 114–15, 119, 135, 160–61, 170–71n3; pollution of, 88, 91, 97, 114, 158–59, 161
ethics across the curriculum, 150
Européenne Ingénieur (Eur Ing), 111–12

familism, 5, 12–13, 20–21, 23, 28, 171n7, 175, 185, 187, 189
Fédération Européenne d'Associations Nationales d'Ingénieurs (FEANI), 111
Fédération Internationale des Ingénieurs-Conseils (FIDIC), 68–70, 72–74, 76, 77n11, 78n14, 78n16, 127–28
Federation of Japan Electric Communications Industrial Association (FJECIA), 47, 57n8, 58n16, 58n18
feudalism, 6, 31n2, 46–47, 75
Fudano Jun, 135, 139, 155, 161–65, 169–70, 172n20
Fujie Kunio, 160–62
Fukushima Daiichi nuclear disaster, 2, 169
Fukuzawa Yukichi, 3
Fundamentals of Engineering (FE), 129
Furuya Keiichi, 160, 172n19

Gendai shisō, 158, 160, 181–82
General Headquarters, Supreme Commander for the Allied Powers (GHQ/SCAP), 3, 35–36, 39, 43, 61, 64, 68, 75, 77n4, 99, 110, 175, 184, 189. *See also* Civil Communications Section; Civil Information and Educational Section; Scientific and Technical Division of the Economic and Scientific Section
gentlemanliness, 3, 12, 14, 16, 32n7, 32n9, 175
Gijutsushi, 1, 6, 61–78, 99–100, 127–46, 152, 167–68, 172n13, 177, 181–83; Act on, 61, 68–70, 72–73, 76, 77n12, 78n13, 127, 130–31, 137, 144n6, 167, 182; Fukumu Yōkō of (1951 code), 2, 61–62, 65–67, 69–71, 75–76, 127, 141, 176, 182; Gyōmu Rinri Yōkō of (1961 code), 2, 61–62, 67–76, 78n13, 127, 136, 141, 182
globalization, 1–2, 5–7, 76, 95, 119, 122, 127, 132, 135, 147, 149–51, 172n13, 175, 177–78; of economy, 5–7, 76, 122, 132, 147, 150, 175, 177; of engineering qualification, 7, 172n13
Great East Japan Earthquake, 2, 169, 172n16
Great Hanshin-Awaji Earthquake, 2, 7n1, 122, 158–59
Gunn, Alastair S., 160

Harada Kosaku, 112–13, 149
Harding, Francis C., 99
Harris, Charles E., Jr., 127, 166, 169
Hastings Center, 4–5
Hazen, Harold L., 51, 57–58n12, 58n20
High Treason Incident, 13
Hippocratic Oath, 98, 115
Hirayama Fukujiro, 13–14, 26, 63–64, 66–69, 71–74, 182
Hitachi, Ltd., 29, 53, 84, 95n3, 99, 140–41, 146n35
HIV-tainted blood scandal, 157–59
Honda Naoshi, 74

Ikeda Hayato, 84–85
Ikeuchi Satoru, 159
Illinois Institute of Technology, 5
Imai Kaneichiro, 100, 112, 125n23, 128–29, 144–45n15, 182–83

Imamichi Tomonobu, 116, 167, 173n27
Imperial College of Engineering, 9–10, 50, 183
Imperial Rescript on Education, 10, 99, 185
Imperial University, 21, 29, 125n20; of Kyoto, 14; of Kyushu, 8n10, 14, 17–18, 29; of Nagoya, 29; of Tohoku, 8n10; of Tokyo, 3, 10, 14–18, 22–23, 32n12, 50, 63–64, 187. *See also* the University of Tokyo
independence: academic, 8n10, 79–80, 85, 87, 89, 92–93, 113, 176; of all commercial interest (and neutrality of the position), 67–76, 77n11, 78n13, 128, 136–37, 182
individualism, 2, 6, 12–13, 19, 23, 32n11, 40, 42, 45, 48, 54, 74, 86, 92, 110, 116–17, 119, 121–22, 134, 145n20, 149, 151, 153–60, 162, 170, 171n7, 172n17, 177–79
industry-academia cooperation (collaboration), 6–7, 8n10, 10, 30, 47, 49–54, 56, 59n23, 79–95, 95–96n5, 96n6, 96n7, 101–3, 106–8, 112–13, 124n18, 151–52, 176, 178, 184; education, 10, 47, 49–54, 59n23, 84, 184; and government collaboration, 101–3, 129, 151–52; and military cooperation, 79, 85. *See also* sandwich program
information ethics, 5, 114, 116, 119, 121–22, 126n33
Information Processing Society of Japan (IPSJ), 2, 7, 117, 119–22, 126n30, 126n33, 127, 151, 183
innovation, 82–84, 102, 104–6, 111, 118, 125n22, 155, 164
Inose Hiroshi, 6, 79, 104–5, 107–8, 123n5, 123–24n10, 124n13, 124n17, 183, 187
Inoue Kowashi, 10–11, 32n5
Inoue Tadashiro, 30, 64, 72, 75
Institute for Engineering Education (IFEE), 51. *See also* Engineering Education Mission to Japan

Institute of Electrical and Electronics Engineers (IEEE), 98–99, 121. *See also* American Institute of Electrical Engineers
Institute of Electrical Communication Engineers of Japan (IECEJ), 49, *50*, 183. *See also* Institute of Electronics, Information and Communication Engineering
Institute of Electrical Engineers of Japan (IEEJ), 2, 49, *50*, 98–99, 119, 123n4
Institute of Electronics, Information and Communication Engineering (IEICE), 2, 121–22, 126n33. *See also* Institute of Electrical Communication Engineers of Japan
Institute of Physical and Chemical Research (RIKEN), 22–23, 33n18, 38, 53, 57n3, 187, 189
Institution of Professional Engineers, Japan (IPEJ), 1, 137. *See also* Japan Consulting Engineers Association
International Federation for Information Processing (IFIP), 120–22, 177
International Labour Organization (ILO), 20, 98, 123n4, 125n22
International Moral Education Congress, 12

Japan Accreditation Board for Engineering Education (JABEE), 1, 49, 112, 130, 134, 137, 145n22, 147–52, 157, 162, 166, 171nn7–8, 177–78, 181, 183–84, 186, 189
Japan Civil Engineering Society (JCES), 1, 2, 3, 7, 9, 25–28, 33n23, 33n25, 35, 62, 73–74, 99, 175, 182. *See also* Japan Society of Civil Engineers
Japan Consulting Engineers Association (JCEA), 1, 2, 6, 13, 26, 30, 55, 61–76, 76–77nn1–3, 77nn11–12, 78n15, 99–100, 122, 127–31, 133–43, 143n1, 144n6, 144n9, 144n13, 145n24, 145n26, 146n29, 146n33, 150, 152, 166, 167, 171n4, 173nn26–27, 176, 182; Technical Translation

Center (TTC) of, 127, 138–43, 145n24, 146n29, 167, 173nn26–27. See also Gijutsushi; Institution of Professional Engineers, Japan
Japan Efficiency Association (JEA), 30
Japanese Society for Engineering Education (JSEE), 4, 36, 49, 52, 54–56, 58–59nn22–23, 79, 83–84, 93–95, 112–13, 134–36, 148–52, 169, 171n8, 177, 183–86, 188–89
Japanese Technicians' Union, 23–24, 26, 28, 30
Japanese Union of Scientists and Engineers (JUSE), 30, 41, 155, 165
Japan Federation of Engineering Societies (JFES), 3, 22, 123n3, 128, 130, 133–36, 143n1, 144n8, 148–49, 151–52, 172n21, 177, 183–84, 186, 189. See also Engineering Society
Japan Medical Association (JMA), 98, 115
Japan PE Council, 129
Japan PE/FE Examiners Council, 129
Japan Productivity Center (JPC), 52–53, 56, 82, 96n6
Japan Science Foundation (JSF), 82–83, 95n5
Japan Society for the Promotion of Science (JSPS), 38, 57n4, 104, 126n33, 187; Committee 149 of, 104, 113, 142, 164, 186; Japan Society for the Promotion of Scientific Research (JSPSR), 38, 57n4
Japan Society of Civil Engineers (JSCE), 1, 2, 33n23, 99–100, 119, 121, 171n4. See also Japan Civil Engineering Society
Japan Society of Mechanical Engineers (JSME), 2, 100, 160–62, 165, 182
Japan Technology Transfer Association (JTTAS), 112, 128–29, 144n6, 152, 183
Japan University Accreditation Association (JUAA), 30, 51–52, 110, 145n16, 147–48, 184
JCO's Tokaimura nuclear fuel-processing plant, 2, 7n4, 157, 168

jitsuyōka (development), 37–39, 57n3

KAKENHI, 103, 140, 149, 164–65, 184
Kanazawa Institute of Technology (KIT), 83, 135, 139, 145n21, 161–66, 172n21, 173n25; Applied Ethics Center for Engineering and Science (ACES) of, 165
Kaneshige Kankuro, 51, 104
Kase Kosaku, 54–55
Kato Hiroyuki, 17
Kawano Yasuo, 72, 78n16
Keidanren, 82, 101, 118, 125n20, 135, 152–54, 172n11, 184–86
Keio University, 29, 47, 82
Keizai Doyukai, 82, 85, 92, 154, 185
Kelly, Harry C., 35, 51, 104
Kihara Hidetoshi, 162
Kihara Junji, 90
Kimura Hisao, 98–99, 123n3
knowledge-based economy, 107, 113, 119, 122, 151, 177
Koga Issac, 4, 36, 43–49, 51–52, 57n12, 124n17, 185
Kokutai (Japan's national polity), 10–11, 13, 16, 19–20, 24, 28, 35, 185. See also emperor
Komatsubara Eitaro, 12
Kōseikai, 22, 30, 33n22, 187
Kudo Hisha, 138–40, 145n25
Ku Klux Klan (KKK), 15

League of Nations, 20, 24
loyalty, 10–13, 16, 25–26, 35, 100, 185. See also Bushido; Confucianism
Luegenbiehl, Heinz C., 163–65

MacArthur, Douglas, 64, 77n7
Martin, Mike W., 116, 139
medical ethics, 98, 114–15, 123n2, 157
Meiji College of Technology (MCT), 14, 16–17, 58n21, 188
Ministry of Education (MOE), 10–12, 16, 32n5, 37–38, 43, 51–53, 58–59nn21–22, 81, 84–85, 103, 107, 110–13, 123n9, 124n15, 125n20,

125n22, 134–35, 137, 147, 152, 184, 188; Higher Education Bureau of, 110, 112, 124n15, 125n22, 152; Science Council of (SCM), 103, 123–24n10; University Council of, 110–11, 125n20. *See also* Ministry of Education, Culture, Sports, Science and Technology

Ministry of Education, Culture, Sports, Science and Technology (MEXT), 145n17, 183. *See also* Ministry of Education

Ministry of Health and Welfare (MOHW), 115, 152, 157

Ministry of Home Affairs, 21–23, 25, 32–33n15, 77n4

Ministry of Industry, 9–10

Ministry of International Trade and Industry (MITI), 52, 63, 68, 73, 103, 106, 108, 135, 140, 143n2, 146n32, 152

Ministry of Posts and Telecommunications (MOPT), 22, 37, 43–44, 152

Miyamoto Takenosuke, 23–26, 28, 74

Monju nuclear reactor, 2, 7n3, 157–60, 177

Morihiro Eiichi, 139, 146n29

Morito Tatsuo, 18, 32n12

Mori Wataru, 108–9

Mukaibo Takashi, 88, 96n8, 104, 186

Murakami Yoichiro, 116–18, 122, 125–26n28, 126n30, 155, 158, 160, 171n5, 172n20, 186–87

Nagaoka Hantaro, 38

Nakasone Yasuhiro, 106, 109, 113, 115

Nanzan University Institute for Social Ethics, 116, 119

Naoki Rintaro, 21–22, 32–33n15

National Academy of Engineering (NAE), 104, 124n12, 158

National Academy of Sciences (NAS), 104, 158, 172n18

National Council of Examiners for Engineering and Surveying (NCEES), 129

nationalism, 3, 13–15, 19, 24–28, 32n11, 33n24, 35, 37, 40, 43, 55–56, 59n27, 99, 124n14, 175–76, 181

National Research Council (NRC), 113

National Science Foundation (NSF), 5, 40, 104, 147

National Society of Professional Engineers (NSPE), 4, 66, 116, 138–39

Nawa Kotaro, 120–22, 126n31

neoliberalism, 152–54, 176

Nihon University, 47, 95n5

Nikkeiren, 80–82, 94, 119, 184–86

Nitobe Inazo, 12, 20, 47, 181

noblesse oblige, 47

North American Free Trade Agreement (NAFTA), 132, 170n1

Official Development Assistance (ODA), 140, 143n4, 146n33

Ohashi Hideo, 136, 148–49, 171n6, 186

Ohnaka Itsuo, 147–48, 150, 170n2

Okamura Sogo, 104–5, 112–13, 123–24n10, 187

Okochi Masatoshi, 22–23, 25, 33n17, 33n19, 38, 57n3, 187

Okoshi Makoto, 52–53, 84, 91

onjō shugi, 20–21, 23–25, 187

Orde van Nederlandse Raadgevende Ingenieurs (ONRI), 70

Organisation for Economic Co-operation and Development (OECD), 40, 57n5, 102, 104, 113, 123n8, 125nn21–22, 183, 186; Committee for Scientific and Technological Policy (CSTP) of, 104, 125n21, 183, 186

Osaka Higher Technical School. *See* technical school

Oyama Matsujiro, 54, 58n22

Physical Society of Japan, 80

pogrom, 18

Index

Polkinghorn, Frank A., 36, 39–49, *50*, 57–58nn12–13, 58n15
Professional Engineer (PE), 1, 54, 61–62, 64–67, 70–72, 95, 112, 127–33, 136–38, 142, 144nn5–9, 144–45n15, 145n22, 152, 167–68, 171n8, 176–78, 182–83. *See also* Institution of Professional Engineers, Japan; National Society of Professional Engineers
Protzman, Charles W., 41–42
public safety, 4, 137, 144–45n15, 168, 170–71n3

RCAST. *See* Research Center for Advanced Science and Technology
Rensselaer Polytechnic Institute, 5
research and development (R&D), 8n10, 36–37, 39–44, *45*, 49, 56, 56n1, 85, 92–93, 101–4, 106–8, 110, 115, 176. *See also* Civil Communications Section
Research Center for Advanced Science and Technology (RCAST), 107–8, 116, 118, 122, 124n18, 125n26, 160, 183, 186, 187
responsibility, 2–3, 5, 16, 41, 46, 47, 55, 56n2, 64, 68, 74, 97–100, 110, 117–18, 120–22, 134, 137, 149–51, 153–56, 158, 160–61, 164, 169–70, 170–71n3, 177–78; individual, 153–56, 158, 177–78; professional, 3, 100, 150; social, 5, 41, 97–99, 117–18, 121, 137, 149–50, 154, 164, 169–70, 178
RIKEN. *See* Institute of Physical and Chemical Research
Rose-Hulman Institute of Technology (RHIT), 163
Russell–Einstein Manifesto, 97
Russo-Japanese War, 12, 16–17, 32n10, 100, 181, 187–88

Saito Tetsuo, 131–32
Sakurai Joji, 38

sandwich program, 9–10, 49–51, 53–54, 79, 83–84, 92–93. *See also* industry-academia cooperation
San Francisco Bay Area Rapid Transit (BART), 4, 98
Sarasohn, Homer M., 41–42, *45*, 46, 57n7
science, technology, and society (STS), 2, 5, 162; STS Network Japan, 162. *See also* science and technology studies
Science and Technology Agency (STA), 56, 68, 77n10, 77n12, 103, 123n1, 131–33, 135, 137, 139, 144n13, 145n16, 146n32, 152, 167–68
science and technology studies, 113, 150, 162, 177, 181–82, 186. *See also* science, technology, and society (STS)
Science Council of Japan (SCJ), 35, 51, 79, 85, 97–99, 101, 112, 115, 117, 123n3, 123n8, 124n12, 130, 135–36, 149, 151, 160, 165, 170–71n3, 171n5, 183, 188–89; Section 5 of, 85, 112, 135, 148, 171n5, 186–88
Science Council of the MOE (SCM). *See* Ministry of Education
Scientific and Technical Division of the Economic and Scientific Section (ESS/ST), 35, 51. *See also* General Headquarters, Supreme Commander for the Allied Powers
Shibata Kiyoshi, 162
Shibusawa Eiichi, 20
Shimizu Kinji, 4, 52, 54–56, 58–59n21–22, 59n26, 85, 188
Shiraishi Tashiro, 63–64, 77n4, 77n7
socialism, 12–13, 18–19, 25, 132
social status: of engineers, 6–7, 9, 21–23, 29–31, 33n22, 64, 66, 69, 72–74, 76, 127–29, 131, 136, 141–43, 144n7, 144n10, 148, 170, 175–78; of scientists, 97
sovereignty, 10–11, 19–20, 28, 32n4, 99

Spencer, Herbert, 15, 17–18, 32n11, 59n27
Standards for Establishment of Universities (SEU), 110–11, 114, 133, 135, 148, 163, 165, 177, 184, 188
student activism, 6, 54, 79, 86, 158, 172n16. *See also* college riots
Sugimoto Taiji, 139, 144n6, 146n28, 167
survival of the fittest, 15, 17, 25
sustainable development, 131, 150, 161, 170–71n3
Suzuka Telecommunications Training School, 43–44, 57n10

Taki Shigeatsu, 130–31, 134, 136, 138–39, 142, 144nn7–9, 145n17, 145–46n27, 167
Tanaka Hiroshi, 64–67, 70–71, 73, 77n8, 127
Taylor, Frederick W., 23, 25, 40–41
technical school, 3, 8n8, 10–11, 13–14, 29, 33n20, 50, 82, 95, 175, 188; Osaka Higher, 10, 13, 50; Tokyo (Higher), 3, 10–11. *See also* Tokyo Institute of Technology
Technical Translation Center (TTC). *See* Japan Consulting Engineers Association (JCEA)
technology ethics (techno-ethics), 115–18, 125n26, 160–61, 173n27, 178
Tejima Seiichi, 3
Tokyo Chamber of Commerce and Industry, 92–93
Tokyo Electric Power Company (TEPCO), 152, 169
Tokyo (Higher) Technical School. *See* technical school
Tokyo Institute of Technology, 3–4, 10, 30, 47, 52, 125n20, 185. *See also* technical school
Tomizu Hirondo, 16, 18
Toyo University, 53–54, 84, 95n3

Uchida Moriya, 144n8, 149, 171n8
Uchida Nobuya, 20
Ui Jun, 91
Umeda Masao, 135–36, 166, 173n26
United Nations, 114, 143n4, 161, 170–71n3, 186; United Nations Educational, Scientific and Cultural Organization (UNESCO), 97–98, 102, 123n4, 123n9, 125n22, 172n20, 186; World Commission on the Ethics of Scientific Knowledge and Technology (COMEST), 172n20, 186
the University of Tokyo, 4, 6, 10, 15, 36, 47, 51–52, 58n22, 79–80, 84, 87–92, 95n4, 101, 104, 107–9, 124n18, 145n19, 160, 172n19, 183, 185–87, 189. *See also* Imperial University; Research Center for Advanced Science and Technology
University Order, 13, 28
Utsumi Kiyoharu, 64, 73–74

Vandegrift, John L., 43–44
Veblen, Thorstein B., 23, 33n19
vertical organization, 159
Vocational School Order, 13, 31n1

Wada Juro, 115
Wada Koroku, 30, 51
Wada Masao, 83, 93–94
Waseda University, 47, 52, 86, 96n7
Watanabe Hiromoto, 3
World Federation of Engineering Organizations (WFEO), 128, 130, 133, 135, 144n15, 147–48, 165, 170–71nn3–4, 177, 183–84, 188
World Medical Association (WMA), 98, 114, 123n2
World War I, 8n10, 13, 20, 22–23
World War II (Pacific War), 3, 7, 9, 13–14, 26, 28, 30–31, 31n1, 35–36, 61–62, 68, 79, 97, 152, 157, 160, 175, 177, 179, 185, 188

Yagi Hidetsugu, 14
Yamakawa Kenjiro, 14–19, 25, 32nn7–9, 33n18
"yellow peril," 18
Yoshida Goro, 37–38, 43, 56–57n2
Yoshida Shigeru, 62, 64, 77nn6–7
Yoshikawa Hiroyuki, 91, 125n21, 136, 149, 151, 160, 189

Yoshino Sakuzo, 19–20
Youmans, Edward L., 15
Young Report, 104

zaibatsu, 23, 44, 75, 189; Furukawa Zaibatsu, 8, 189. *See also* Institute of Physical and Chemical Research (RIKEN)

About the Author

Natsume Kenichi is an associate professor at the Humanities and Social Sciences Program, Kanazawa Institute of Technology, where he teaches engineering ethics and conducts research on science, technology, and society (STS) from a historical perspective. In addition to the papers on engineering ethics noted in this book, he published a paper on Japan's dual-use technology research and foreign affairs and co-authored several engineering ethics textbooks. Natsume is a board member of the History of Science Society of Japan, the Japanese Society for Science and Technology Studies, and the Japan Society for the History of Industrial Technology. He has served as the chairperson of the Research Ethics Committee of the History of Science Society of Japan and is also a member of the Engineering Ethics Investigation and Research Committee of the Japanese Society for Engineering Education.